The Unofficial   **Ages 8+**

MINECRAFT
MATH WORKBOOK

FRACTIONS

- ✓ Coloring
- ✓ Tricks
- ✓ Mazes
- ✓ Word Search
- ✓ Puzzles
- ✓ Word problems

First published in the USA 2019.   ISBN 9781948737517

Contents

I'm highly confident, adventurous, and open-minded. I'm ready to help or take any risk. I'm an encouraging and unbelievably good friend.

I'm the grumpy, bad-tempered, critical Brainer. I get "crazy-mad" and impatient with "any problem." I would be the happiest-ever-lived Brainer if not for math.

I'm "the-smartest-ever-lived Brainer;" ready to solve any problem; and happy to explain anything to any Brainer. I'm hard-working and brainy.

I'm scared of everything, especially new stuff. I get panicky and terrified by many pages of a workbook.

I'm enthusiastic, excited, and sure about everything in this beautiful world! I'm friendly, kind, and wise!

I love to be left alone. I would be a dreamer if not for other Brainers and math. Sometimes I'm ready to give up right away or resist anything new…"WHY should I???" I'm persistent in doing nothing.

1. <u>Read and write</u> the missing numbers.

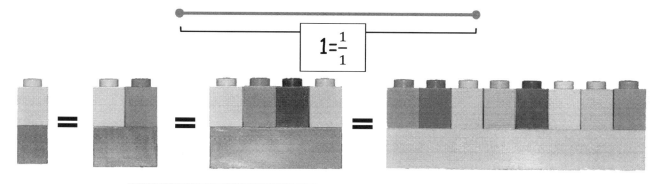

$$1 = \frac{1}{1}$$

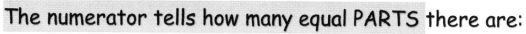

The numerator tells how many equal PARTS there are:

$$\text{Whole} = 1 = \frac{2}{2} = \frac{4}{4} = \frac{8}{8}$$

The denominator tells the number of equal parts in the WHOLE:

$$\frac{1}{2}$$

A half = $\frac{1}{2}$.

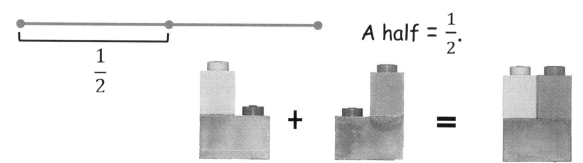

I have 1 blue and 1 orange parts out of 2 equal parts:

$$\frac{1}{2} + \frac{1}{2} = \frac{\ldots + \ldots}{2} = \frac{2}{2} = \ldots$$

The whole is 2 (has 2 equal parts).

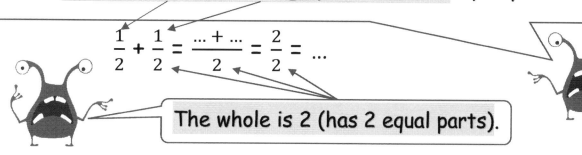

1. <u>Answer</u> the questions and <u>write</u> the missing numbers.

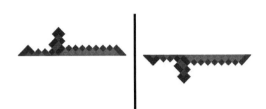

<u>Divide</u> the sword equally into 2 parts.

$1 \div 2 = \frac{1}{2}$

It's divisible (I can divide it).

To divide is to split a sword into equal parts. I write the division sign: "÷" to show division or I use the fractional line to show division: —.

1 sword is divided into 2 equal parts:

$1 \div 2 = \frac{1}{2}$ (a half).

Take a ruler and <u>divide</u> the rectangle into 2 equal parts by drawing a line.

$... \div ... = \frac{...}{...}$

Take a ruler and <u>divide</u> the block into 2 equal parts by drawing a line.

$... \div ... = \frac{...}{...}$

www.stemmindset.com

1. <u>Answer</u> the questions and <u>write</u> the missing numbers.

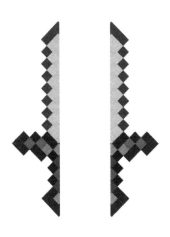

I have ⬚2⬚ ⬚halves⬚ ($\frac{1}{2}$) of a sword.

<u>How many</u> do I have in all?

$$\frac{1}{2} + \frac{1}{2} = \frac{\dots + \dots}{\dots} = \frac{\dots}{\dots} = \dots \text{ whole sword.}$$

Look, these are like fractions since they all have the like denominator: 2.

So, you can write ⬚2⬚ as a common denominator for both fractions.

And then, add the numerators 1 and 1.

I have ⬚1⬚ whole sword and ⬚a half⬚ ($\frac{1}{2}$) of a sword.

<u>How many</u> do I have in all?

$$1 + \frac{1}{2} = \frac{2}{2} + \frac{\dots}{\dots} = \frac{\dots + \dots}{\dots} = \frac{\dots}{\dots} = \dots \frac{\dots}{\dots}$$

or 1 whole sword and a half.

www.stemmindset.com

1. <u>Answer</u> the questions and <u>write</u> the missing numbers.

I have ☐1 whole sword and ☐a half ($\frac{1}{2}$) of a sword. If I add a half ($\frac{1}{2}$) of a sword <u>how many</u> do I have in all? (You can choose which algorithm you like the most).

$$1\frac{1}{2} + \frac{1}{2} = \frac{3}{2} + \frac{1}{2} = \frac{...+...}{...} = \frac{...}{...} = ... \text{ swords}$$

Add fractions   or   Add wholes

$$1\frac{1}{2} + \frac{1}{2} = ... \qquad 1)\ \frac{1}{2} + \frac{1}{2} = \frac{...+...}{...} = \frac{...}{...} = ... \qquad 2)\ 1 + 0 = ...$$

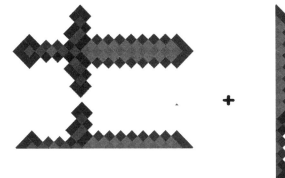  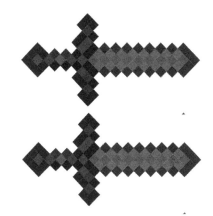

2. On Monday I get my homework for the week, usually I do $\frac{2}{8}$ of my homework per day. <u>How much of my homework</u> will be done in 2 days?

Per day	Days	Homework
$\frac{...}{...}$	...	$?\ \frac{...}{...}$

$$1 = \frac{...}{...}$$

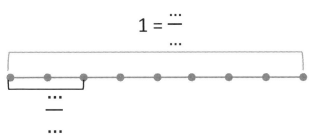

Answer:____

1. **Answer** the questions and **write** the missing numbers and words.

I had a block. I shared a half with my friend. **What part** is left?

$$1 - \frac{1}{2} = \frac{...}{...} - \frac{...}{...} = \frac{... - ...}{...} = \frac{...}{...} \;(\underline{\hspace{6cm}}).$$

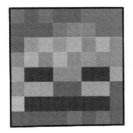  

2. **Subtract** the fractions. The first one is done for you.

$$\frac{3}{4} - \frac{1}{4} = \frac{3 - 1}{4} = \frac{2}{4}$$

$$\frac{3}{4} - \frac{2}{4} = \frac{... - ...}{...} = \frac{...}{...}$$

$$\frac{6}{8} - \frac{3}{8} = \frac{... - ...}{...} = \frac{...}{...}$$

$$\frac{3}{8} - \frac{1}{8} = \frac{... - ...}{...} = \frac{...}{...}$$

$$\frac{7}{8} - \frac{5}{8} = \frac{... - ...}{...} = \frac{...}{...}$$

$$\frac{5}{8} - \frac{1}{8} = \frac{... - ...}{...} = \frac{...}{...}$$

 www.stemmindset.com

1. <u>Answer</u> the questions. <u>Write</u> the missing numbers and words.

I had $\boxed{1}$ whole block and a $\boxed{\text{half}}$ $(\frac{1}{2})$ of a block.

I shared $\frac{1}{2}$ of a block with a friend. <u>What part</u> is left?

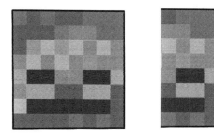

You can <u>choose</u> one of the two strategies:

$1\frac{1}{2} - \frac{1}{2} = \frac{...}{...} - \frac{...}{...} = \frac{...-...}{...} = \frac{...}{...}$ or _____.

or

$1\frac{1}{2} - \frac{1}{2} = ...$ 1) $\frac{...}{...} - \frac{...}{...} = \frac{...-...}{...} = ...$ 2) $... - ... = ...$

2. <u>Write</u> the missing numbers. <u>Add</u> and <u>compare</u> the fractions using ">" or "<."

$\frac{1}{6} + \frac{1}{6} = \frac{...+...}{...} = \frac{...}{...}$ ... $\frac{3}{6} + \frac{2}{6} = \frac{...+...}{...} = \frac{...}{...}$

1. <u>Find</u> and <u>circle</u> or <u>cross out</u> the words to find out more about Minecraft.

U M B C S K E C Y M E H
L N J E N P O E E I H F   CORRIDOR
B I D A W R P J C N T U
S X L E R B S M N E K W   COBWEB
W P F I R T O J E S W Z   MINESHAFT
I C D S O G O C F H W S
Z O F L O O R Q Q A C Y   FENCE
R H K V T B W O H F Y U   UNDERGROUND
G N I L I E C X U T X F
E N S C Z Z Z N Z N E C   PLANK
H O H P Q J R K M G D B   CEILING
O O N T Z A K O L D R K   FLOOR

www.stemmindset.com

1. <u>Read and write</u> the missing numbers.

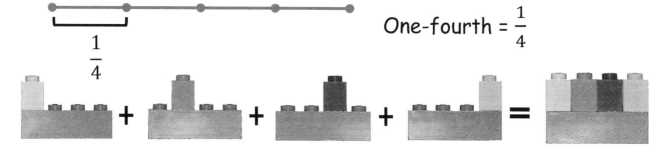

One-fourth = $\frac{1}{4}$

I added four one fourths. They are like fractions (fractions with like denominators).

$$\frac{1}{4} + \frac{1}{4} + \frac{1}{4} + \frac{1}{4} = \frac{... + ... + ... + ...}{4} = \frac{...}{4} = ...$$

The whole is made of 4 equal parts. Done!

One-eighth = $\frac{1}{8}$

$\frac{1}{8}$

I added eight one eighths.

$$\frac{1}{8} + \frac{1}{8} + \frac{1}{8} + \frac{1}{8} + \frac{1}{8} + \frac{1}{8} + \frac{1}{8} + \frac{1}{8} = \frac{... + ... + ... + ... + ... + ... + ... + ...}{8} = \frac{...}{8} = ...$$

1.   <u>Fill in</u> the missing numbers.

equals $\dfrac{1}{2}$ as

equals $\dfrac{...}{...}$

equals $\dfrac{3}{3}$ as

equals $\dfrac{...}{...}$

2.   My watering can is $\dfrac{1}{4}$ of a bucket. <u>How many watering cans</u> do I need for a whole bucket ?

Can	Bucket	Cans
$\dfrac{...}{...}$	...	? ...

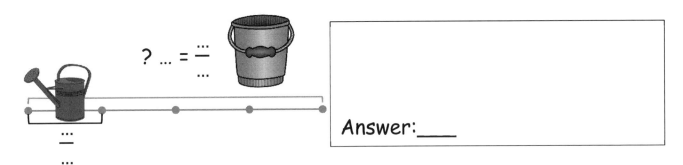

? ... = $\dfrac{...}{...}$

$\dfrac{...}{...}$

Answer:____

3. <u>Write</u> the missing numbers and <u>subtract</u> the fractions.

 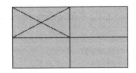 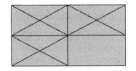 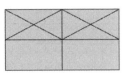

$\dfrac{3}{4} - \dfrac{1}{4} = \dfrac{...}{...}$ $\dfrac{...}{...} - \dfrac{...}{...} = \dfrac{...}{...}$ $\dfrac{...}{...} - \dfrac{...}{...} = \dfrac{...}{...}$ $\dfrac{...}{...} - \dfrac{...}{...} = \dfrac{...}{...}$

                   www.stemmindset.com

1. <u>Compare</u> the fractions using ">," "<," or "=."

$\frac{1}{3}$ ... $\frac{4}{3}$ $\frac{3}{3}$ ... $\frac{2}{3}$ $\frac{2}{3}$ ... $\frac{1}{3}$

$\frac{4}{3}$ ... $\frac{2}{3}$ $\frac{3}{3}$ ... 1 $\frac{2}{2}$ ... $\frac{3}{3}$

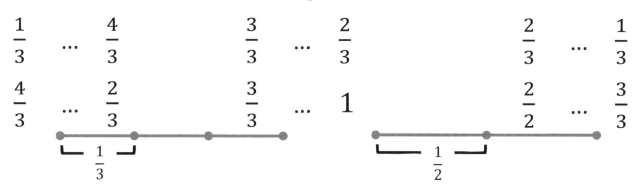

2. <u>Answer</u> the questions.

<u>How many parts</u> are these shapes divided into?   ... parts.

<u>Are</u> these parts equal?   _____.

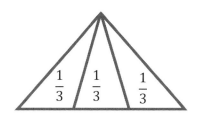

<u>Color</u> $\frac{3}{3}$ red.   <u>Color</u> $\frac{2}{3}$ purple.

| $\frac{1}{3}$ |
| $\frac{1}{3}$ |
| $\frac{1}{3}$ |

<u>Color</u> $\frac{1}{3}$ green.

3. <u>Write</u> the missing numbers and <u>add</u> the fractions.

$\frac{1}{4} + \frac{1}{4} = \frac{\ }{\ }$ $\frac{\ }{\ } + \frac{\ }{\ } + \frac{\ }{\ } = \frac{\ }{\ }$ $\frac{\ }{\ } + \frac{\ }{\ } + \frac{\ }{\ } + \frac{\ }{\ } = \frac{\ }{\ }$

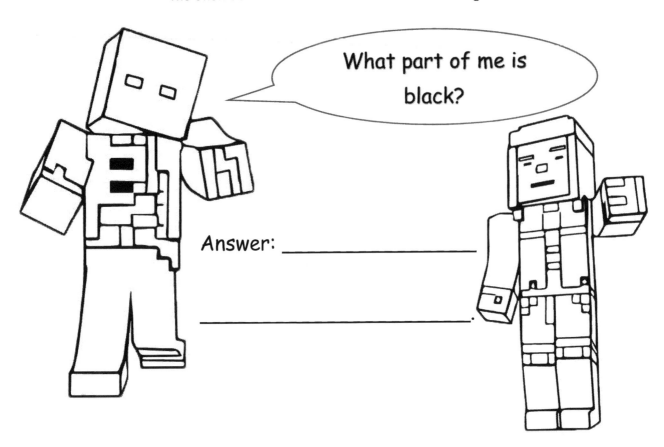

Answer: _____

_____.

1. <u>Answer</u> the question.

A	B	C	D	E	F	G	H	I	J	K	L	M
1	...	...	...	...	...	...	...	...	...	...	...	...

N	O	P	Q	R	S	T	U	V	W	X	Y	Z
...	...	...	...	...	...	...	...	...	...	...	...	26

‾‾ ‾ ‾ ‾‾   ‾ ‾‾   ‾‾ ‾‾ ‾‾   ‾‾ ‾ ‾ ‾   ‾‾ ‾‾
23 8 1 20   4 15   25 15 21   14 5 5 4   20 15

‾ ‾‾ ‾ ‾ ‾‾   ‾   ‾ ‾ ‾ ‾ ‾‾ ‾‾?
3 18 1 6 20   1   2 5 1 3 15 14

_____.

1. <u>Add or subtract</u> the fractions. <u>Draw</u> the arrows to match the equation to the picture and to the answer.

$$\frac{1}{3} + \frac{1}{3} \qquad \frac{2}{3} + \frac{1}{3} \qquad \frac{3}{3} - \frac{1}{3} \qquad \frac{3}{3} - \frac{2}{3} \qquad \frac{2}{3} - \frac{1}{3}$$

$$\frac{1}{3} \qquad\qquad \frac{2}{3} \qquad\qquad \frac{3}{3} \qquad\qquad \frac{1}{3} \qquad\qquad \frac{2}{3}$$

2. <u>Draw</u> the red dot on the number line to answer each question.

<u>Where</u> is $1\frac{1}{2}$?

<u>Where</u> is $1\frac{2}{4}$?

<u>Where</u> is $2\frac{3}{4}$?

1. <u>Compare</u> the fractions using ">," "<," or "=."

$$\frac{1}{3} \quad \dots \quad \frac{2}{3} \qquad\qquad \frac{3}{3} \quad \dots \quad \frac{2}{3} \qquad\qquad \frac{3}{3} \quad \dots \quad 1$$

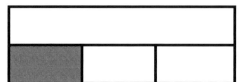

2. <u>Subtract</u> the fractions.

$$\frac{7}{8} - \frac{3}{8} = \frac{\dots - \dots}{\dots} = \frac{\dots}{\dots} \qquad\qquad \frac{7}{8} - \frac{1}{8} = \frac{\dots - \dots}{\dots} = \frac{\dots}{\dots}$$

$$\frac{4}{8} - \frac{2}{8} = \frac{\dots - \dots}{\dots} = \frac{\dots}{\dots} \qquad\qquad \frac{5}{6} - \frac{3}{6} = \frac{\dots - \dots}{\dots} = \frac{\dots}{\dots}$$

$$\frac{4}{6} - \frac{1}{6} = \frac{\dots - \dots}{\dots} = \frac{\dots}{\dots} \qquad\qquad \frac{5}{6} - \frac{1}{6} = \frac{\dots - \dots}{\dots} = \frac{\dots}{\dots}$$

 www.stemmindset.com

1. <u>Answer</u> the questions and <u>write</u> the missing numbers.

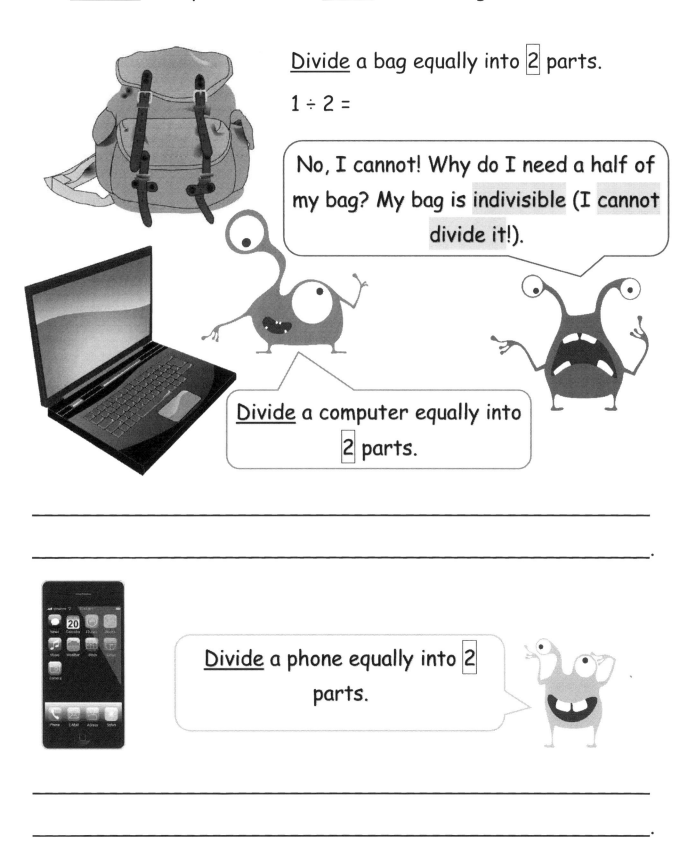

<u>Divide</u> a bag equally into 2 parts.

1 ÷ 2 =

No, I cannot! Why do I need a half of my bag? My bag is indivisible (I cannot divide it!).

<u>Divide</u> a computer equally into 2 parts.

_____

_____.

<u>Divide</u> a phone equally into 2 parts.

_____

_____.

1. What are you thinking of?

Which is longer, one fifth or two fourths?

Answer: _____

_____.

1. <u>Color</u> $\frac{1}{2}$ *blue,* $\frac{2}{3}$ *yellow,* $\frac{2}{4}$ *red,* $\frac{5}{8}$ *green.*

1

	$^1/_2$

		$^1/_3$

			$^1/_4$

							$^1/_8$

2. <u>Compare</u> the fractions using ">," "<," or "=." It may be helpful to look at the fractions strips above.

$\frac{1}{3}$ ... $\frac{1}{8}$ $\frac{1}{3}$ ... $\frac{3}{8}$

$\frac{1}{2}$ ... $\frac{2}{4}$ $\frac{1}{3}$ ... $\frac{1}{4}$

$\frac{1}{3}$ ... $\frac{1}{2}$ $\frac{3}{4}$ ... $\frac{6}{8}$

$\frac{3}{3}$ ... $\frac{1}{2}$ $\frac{2}{2}$ ... $\frac{7}{8}$

$\frac{2}{4}$ ... $\frac{6}{8}$ $\frac{2}{4}$ ... $\frac{2}{8}$

$\frac{2}{4}$ ... $\frac{4}{8}$ $\frac{2}{3}$ ... $\frac{3}{4}$

1.   <u>Write</u> the missing numbers and <u>subtract</u> the fractions. The first one is done for you.

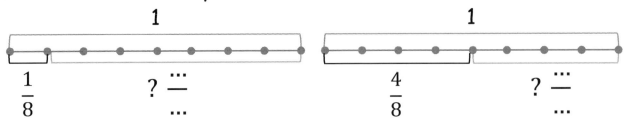

$\dfrac{1}{8}$ $? \dfrac{...}{...}$ $\dfrac{4}{8}$ $? \dfrac{...}{...}$

$1 - \dfrac{1}{8} = \dfrac{8}{8} - \dfrac{...}{...} = \dfrac{... - ...}{...} = \dfrac{...}{...}$ $1 - \dfrac{4}{8} = \dfrac{...}{...} - \dfrac{...}{...} = \dfrac{... - ...}{...} = \dfrac{...}{...}$

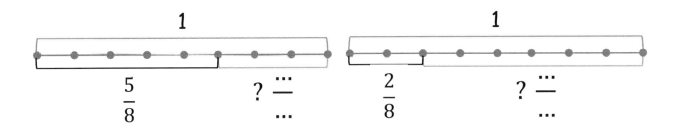

$\dfrac{5}{8}$ $? \dfrac{...}{...}$ $\dfrac{2}{8}$ $? \dfrac{...}{...}$

$1 - \dfrac{5}{8} = \dfrac{...}{...} - \dfrac{...}{...} = \dfrac{... - ...}{...} = \dfrac{...}{...}$ $1 - \dfrac{2}{8} = \dfrac{...}{...} - \dfrac{...}{...} = \dfrac{... - ...}{...} = \dfrac{...}{...}$

2.   <u>Add</u> the fractions and <u>write</u> the value in the rectangles below.

$\dfrac{1}{6} + \dfrac{1}{6}$ $\dfrac{2}{6} + \dfrac{1}{6}$ $\dfrac{2}{6} + \dfrac{2}{6}$ $\dfrac{3}{6} + \dfrac{3}{6}$ $\dfrac{4}{6} + \dfrac{2}{6}$ $\dfrac{5}{6} + \dfrac{1}{6}$

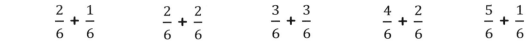

$\dfrac{\Box}{6}$ $\dfrac{\Box}{6}$ $\dfrac{\Box}{6}$ $\dfrac{\Box}{6}$ $\dfrac{\Box}{6}$ $\dfrac{\Box}{6}$

1. <u>Write</u> the missing numbers and <u>add</u> the fractions.

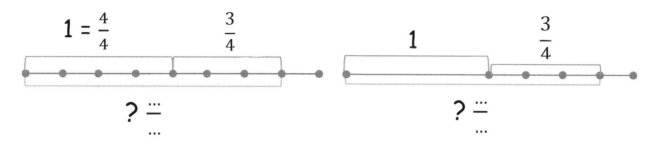

$1 + \dfrac{3}{4} = -\!-\!+\!-\!- = $ ——————— $ = -$

$1 + \dfrac{3}{4} = -\!-\!+\!-\!- = $ ——————— $ = -$

2. My sister has ③ hats and she wants to share them equally with her friend. <u>Draw</u> ③ hats and <u>answer</u> <u>how</u> they can divide the hats equally.

Answer: _____.

1. Tell me, how many equal parts of the whole are in six eights? Four tenths?

Answer: _____

_____.

2. <u>Unscramble</u> each word to find out more about Minecraft. <u>Write</u> the word in the squares.

sirla

orin

hsects

nearsmict

eropxel

www.stemmindset.com

1. <u>Subtract</u> the two fractions. <u>Write</u> the value in the box below.

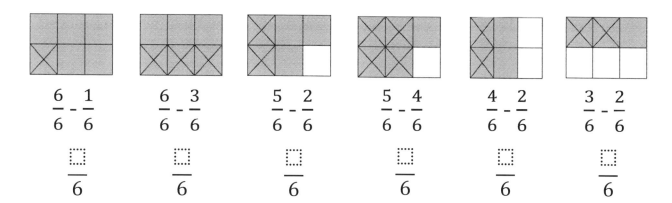

$$\frac{6}{6} - \frac{1}{6} \qquad \frac{6}{6} - \frac{3}{6} \qquad \frac{5}{6} - \frac{2}{6} \qquad \frac{5}{6} - \frac{4}{6} \qquad \frac{4}{6} - \frac{2}{6} \qquad \frac{3}{6} - \frac{2}{6}$$

$$\frac{\boxed{}}{6} \qquad \frac{\boxed{}}{6} \qquad \frac{\boxed{}}{6} \qquad \frac{\boxed{}}{6} \qquad \frac{\boxed{}}{6} \qquad \frac{\boxed{}}{6}$$

2. <u>How many halves</u> ($\frac{1}{2}$) are in $\boxed{1 \text{ whole}}$? <u>Color</u> one half green and another half yellow.

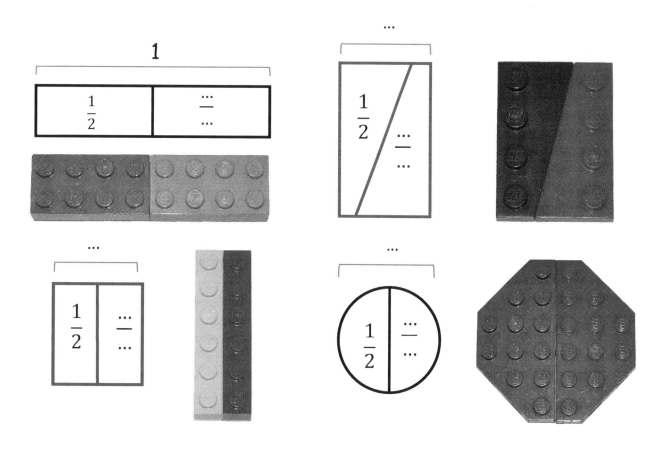

1. <u>Subtract</u> the fractions. <u>Color</u> the minuend yellow and <u>cross out</u> the subtrahend for each problem.

$$\frac{6}{6} - \frac{3}{6} \qquad \frac{6}{6} - \frac{1}{6} \qquad \frac{5}{6} - \frac{4}{6} \qquad \frac{4}{6} - \frac{1}{6} \qquad \frac{3}{6} - \frac{2}{6} \qquad \frac{2}{6} - \frac{1}{6}$$

$$\frac{\square}{6} \qquad\qquad \frac{\square}{6} \qquad\qquad \frac{\square}{6} \qquad\qquad \frac{\square}{6} \qquad\qquad \frac{\square}{6} \qquad\qquad \frac{\square}{6}$$

2. <u>Subtract</u> the fractions.

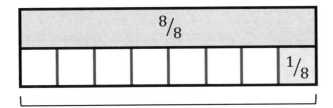

$$\frac{3}{8} - \frac{1}{8} = \frac{...-...}{...} = \frac{...}{...} \qquad\qquad \frac{8}{8} - \frac{4}{8} = \frac{...-...}{...} = \frac{...}{...}$$

$$\frac{7}{8} - \frac{5}{8} = \frac{...-...}{...} = \frac{...}{...} \qquad\qquad \frac{6}{8} - \frac{1}{8} = \frac{...-...}{...} = \frac{...}{...}$$

www.stemmindset.com

1. <u>Answer</u> the questions and <u>fill in</u> the missing numbers.

A 5-inch long rectangle is <u>divided</u> into 5 equal parts.

> If I need 5 equal parts, I always draw a 5-inch long rectangle and cut it into 5 equal 1-inch long parts. Then, I get 5 equal parts.

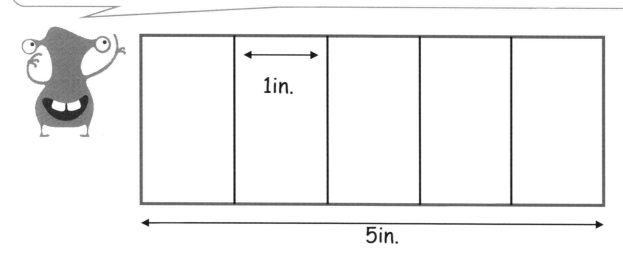

5in.

<u>What fraction names</u> 1 part out of 5? $1 \div 5 = \frac{1}{5}$ (one fifth).

<u>What fraction names</u> 2 parts out of 5? $... \div ... = \frac{...}{...}$ (_____).

<u>What fraction names</u> 3 parts out of 5? $... \div ... = \frac{...}{...}$ (_____).

<u>What fraction names</u> 4 parts out of 5? $... \div ... = \frac{...}{...}$ (_____).

<u>What fraction names</u> 5 parts out of 5? $... \div ... = \frac{...}{...} = 1$.

> The whole always equals
> 1 or $\frac{2}{2}$, or $\frac{3}{3}$, or $\frac{4}{4}$, or $\frac{5}{5}$.

1. Anjkdihfjskl iwbblblbgs nicsobdobo.

Write the name for just one part of seven tenths.

_____

2. Find and circle or cross out the words to find out more about anvils in Minecraft.

```
Y  X  M  T  L  P  P  E  K  J  R  R
C  C  T  D  O  F  N  U  R  P  N  E
C  G  N  W  O  I  X  R  V  A  S  N
A  X  E  E  B  R  I  W  M  W  S  A
H  R  P  M  I  N  J  B  U  S  E  M
N  Q  O  D  S  C  Q  E  W  X  N  E
R  C  G  R  A  V  I  T  Y  F  P  P
H  E  T  I  M  S  N  F  H  C  R  V
Z  T  P  H  T  Y  V  H  F  I  A  Y
Z  N  T  A  P  M  A  M  Y  E  H  F
Z  O  D  S  I  L  J  I  Q  K  S  O
U  F  Y  I  I  R  C  Y  V  Y  M  P
```

REPAIR

RENAME

COMBINE

EFFICIENCY

SHARPNESS

POWER

SMITE

GRAVITY

www.stemmindset.com

1. <u>Add</u> the fractions. <u>Color</u> the squares red for the first addend and blue for the second addend. <u>Write</u> the value in the box below.

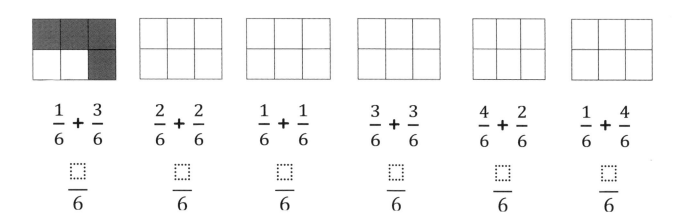

$\dfrac{1}{6} + \dfrac{3}{6}$ $\dfrac{2}{6} + \dfrac{2}{6}$ $\dfrac{1}{6} + \dfrac{1}{6}$ $\dfrac{3}{6} + \dfrac{3}{6}$ $\dfrac{4}{6} + \dfrac{2}{6}$ $\dfrac{1}{6} + \dfrac{4}{6}$

$\dfrac{\square}{6}$ $\dfrac{\square}{6}$ $\dfrac{\square}{6}$ $\dfrac{\square}{6}$ $\dfrac{\square}{6}$ $\dfrac{\square}{6}$

2. I have $\boxed{20}$ wood bricks. $\dfrac{3}{8}$ of them are cubes, $\dfrac{1}{8}$ are cylinders. <u>How many cubes and cylinders</u> do I have? <u>What fraction names</u> pyramids?

Bricks	Cubes	Cylinders	Cubes + cylinders	Pyramids
...	$\dfrac{...}{...}$	$\dfrac{...}{...}$	? ...	? ...

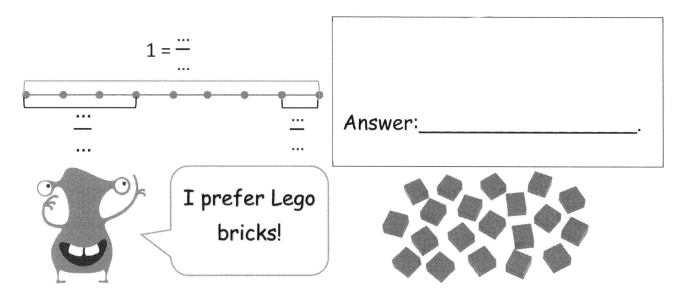

$1 = \dfrac{...}{...}$

$\dfrac{...}{...}$ $\dfrac{...}{...}$

Answer:_____.

I prefer Lego bricks!

Let's look at these triangles. Aha...

We have even numbers – 4, 6, 8, and 10 triangles. Even numbers are numbers that can be divided (split) equally into 2 parts or halves.

$$1 \text{ whole} = \frac{4}{4} = \frac{4 \text{ equal parts}}{4 \text{ in the whole}}$$

Aha... I have 4 triangles - 2 parts out of 4 is a half of 4.

$2 + 2$:

$$\frac{2}{4} = \frac{1}{2} \qquad \frac{2}{4} = \frac{1}{2}$$

And I have 6 triangles - 3 triangles out of 6 is a half of 6 triangles.

$$1 \text{ whole} = \frac{6}{6} = \frac{6 \text{ equal parts}}{6 \text{ in the whole}}$$

$3 + 3$:

$$\frac{3}{6} = \frac{1}{2} \qquad \frac{3}{6} = \frac{1}{2}$$

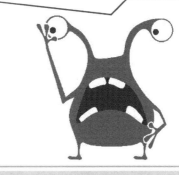

So, I have 8 triangles - 4 triangles out of 8 is a half of 8 triangles.

$$1 \text{ whole} = \frac{8}{8} = \frac{8 \text{ equal parts}}{8 \text{ in the whole}}$$

$4 + 4$:

$$\frac{4}{8} = \frac{1}{2} \qquad \frac{4}{8} = \frac{1}{2}$$

www.stemmindset.com

Look at these 10 triangles - 5 parts out of 10 is a half of 10 triangles.

$$1 \text{ whole} = \frac{10}{10} = \frac{10 \text{ equal parts}}{10 \text{ in the whole}}$$

5 + 5:

$$\frac{5}{10} = \frac{1}{2} \qquad \frac{5}{10} = \frac{1}{2}$$

1. I've read $\frac{7}{8}$ of the book in 2 days. I read $\frac{3}{8}$ on Thursday. <u>How many</u> did I read on Friday?

Read	Thurs.	Fri.
$\frac{...}{...}$	$\frac{...}{...}$	? $\frac{...}{...}$

$$1 = \frac{...}{...}$$

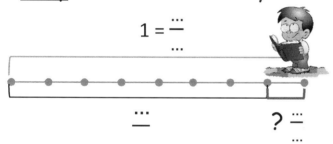

$$\frac{...}{...} \qquad ? \frac{...}{...}$$

Answer:_____

2. I've put 9 blocks equally into 3 boxes. <u>What fraction names</u> 1 box?

Blocks	Boxes	Fraction
...	...	? $\frac{...}{...}$

$$1 = \frac{...}{...}$$

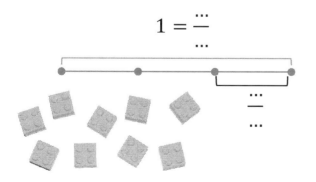

$$\frac{...}{...}$$

Answer:_____

1. <u>Subtract</u> the fractions.

$\dfrac{7}{8} - \dfrac{3}{8} = \dfrac{...-...}{...} = \dfrac{...}{...}$

$\dfrac{7}{8} - \dfrac{1}{8} = \dfrac{...-...}{...} = \dfrac{...}{...}$

$\dfrac{4}{8} - \dfrac{2}{8} = \dfrac{...-...}{...} = \dfrac{...}{...}$

$\dfrac{5}{6} - \dfrac{3}{6} = \dfrac{...-...}{...} = \dfrac{...}{...}$

$\dfrac{4}{6} - \dfrac{1}{6} = \dfrac{...-...}{...} = \dfrac{...}{...}$

$\dfrac{5}{6} - \dfrac{1}{6} = \dfrac{...-...}{...} = \dfrac{...}{...}$

2. I've done $\dfrac{3}{4}$ of my homework. $\dfrac{1}{4}$ was writing a story. <u>What fraction names</u> was math? <u>How much of my homework</u> is left to be done?

Done	Writing	Math	Left
$\dfrac{...}{...}$	$\dfrac{...}{...}$	$? \dfrac{...}{...}$	$? \dfrac{...}{...}$

? ...

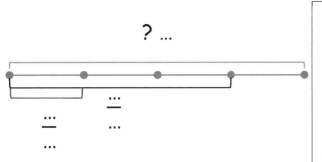

$\dfrac{...}{...}$

$\dfrac{...}{...}$

Answer:_____.

1. How many quarters (one fourths) make a half?

Wow! That's tough. Any ideas, bud?

_____

2. <u>Unscramble</u> each word to find out what armor can protect you. <u>Write</u> the word in the squares.

rworas

slivan

aavl

illbfaser

odsrw

1. <u>Write</u> the missing numbers to make the fractions complete.

Let's look at these triang...

Yeah, I know! I have 3, 6, and 9 triangles. They can be divided equally into 3 parts!

Aha, I have $\boxed{3}$ triangles - 1 triangle out of 3 is one third of 3.

$1 + 1 + 1:$ 1 whole $= \dfrac{3}{3} = \dfrac{...\ equal\ parts}{...\ in\ the\ whole}$

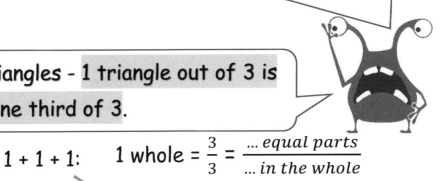

$\dfrac{1}{3}$ $\dfrac{\square}{\square}$ $\dfrac{\square}{\square}$

I have $\boxed{6}$ triangles - 2 triangles out of 6 is one third of 6.

1 whole $= \dfrac{6}{6} = \dfrac{...\ equal\ parts}{...\ in\ the\ whole}$

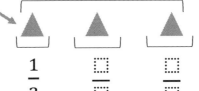

$2 + 2 + 2$

$\dfrac{2}{6} = \dfrac{1}{3}$ $\dfrac{\square}{\square} = \dfrac{\square}{\square}$ $\dfrac{\square}{\square} = \dfrac{\square}{\square}$

Now, I have $\boxed{9\ triangles}$ - 3 parts out of 9 is...

1 whole $= \dfrac{9}{9} = \dfrac{...\ equal\ parts}{...\ in\ the\ whole}$

$3 + 3 + 3:$

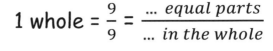

...one third of 9.

$\dfrac{3}{9} = \dfrac{1}{3}$ $\dfrac{\square}{\square} = \dfrac{\square}{\square}$ $\dfrac{\square}{\square} = \dfrac{\square}{\square}$

1. <u>Answer</u> the questions and <u>write</u> the missing numbers.

A 3-inch long square is divided into 3 equal parts each 1-inch long.

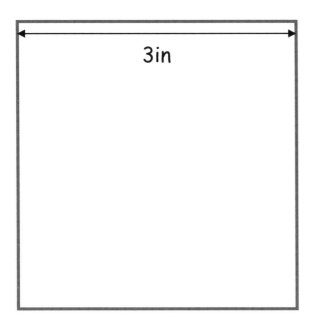

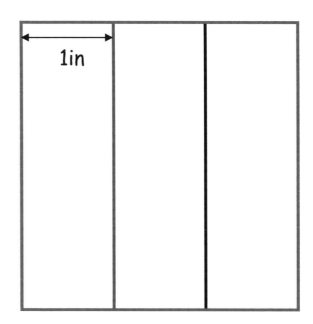

<u>What fraction names</u> 1 part out of 3? $1 \div 3 = \frac{1}{3}$ (one third).

<u>What fraction names</u> 2 parts out of 3? $\ldots \div \ldots = \frac{\ldots}{\ldots}$

(_____).

<u>What fraction names</u> 3 parts out of 3? $\ldots \div \ldots = \frac{\ldots}{\ldots} = 1$

(_____).

The whole always equals 1 or $\frac{2}{2}$, or $\frac{3}{3}$.

1. I've had $\boxed{5}$ apples equally for lunch in $\boxed{5}$ days. <u>What part of the</u> <u>apples</u> have I eaten in $\boxed{\text{three}}$ days?

Apples	Days	Part
...	...	? $\frac{...}{...}$

$$1 = \frac{...}{...}$$

? $\frac{...}{...}$

Answer:_____

2. <u>Add</u> the fractions.

$$\frac{1}{8} + \frac{1}{8} = \frac{... + ...}{...} = \frac{...}{...}$$

$$\frac{5}{8} + \frac{1}{8} = \frac{... + ...}{...} = \frac{...}{...}$$

$$\frac{1}{8} + \frac{3}{8} = \frac{... + ...}{...} = \frac{...}{...}$$

$$\frac{2}{8} + \frac{3}{8} = \frac{... + ...}{...} = \frac{...}{...}$$

$$\frac{4}{8} + \frac{4}{8} = \frac{... + ...}{...} = \frac{...}{...} = ...$$

$$\frac{5}{8} + \frac{3}{8} = \frac{... + ...}{...} = \frac{...}{...} = ...$$

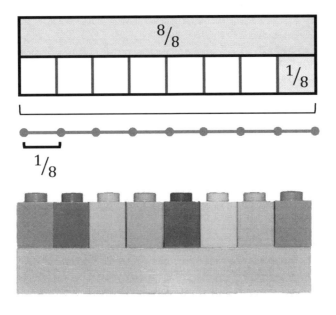

www.stemmindset.com

1. I'm less than 4 wholes.
I'm made of 5 halves.
Who am I?

Easy! I'm ... Hm.... I desperately need your help, BFF!

_____

2. <u>Answer</u> the question.

A	B	C	D	E	F	G	H	I	J	K	L	M
1	...	...	...	...	...	...	...	...	...	...	...	...
N	O	P	Q	R	S	T	U	V	W	X	Y	Z
...	...	...	...	...	...	...	...	...	...	...	...	26

‾ ‾ ‾ ‾   ‾ ‾   ‾ ‾ ‾   ‾ ‾ ‾ ‾   ‾ ‾
23 8 1 20 4 15   25 15 21   14 5 5 4   20 15

‾ ‾ ‾ ‾ ‾   ‾ ‾ ‾ ‾   ‾ ‾   ‾ ‾ ‾ ‾ ‾?
3 18 1 6 20   1 19 5 20   15 6   1 18 13 15 18

_____.

1. <u>What fraction</u> is more? Use ">," "<," or "=."

$\frac{1}{2}$ ... $\frac{1}{4}$ $\frac{1}{8}$ ... $\frac{3}{4}$

$\frac{1}{8}$ ... $\frac{1}{2}$ $\frac{1}{2}$ ... $\frac{2}{4}$

$\frac{3}{4}$ ... $\frac{7}{8}$ $\frac{5}{8}$ ... $\frac{1}{2}$

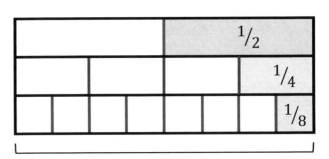

<u>Why</u> is $\frac{1}{2}$ equal to $\frac{2}{4}$? _____

_____.

2. <u>Answer</u> the questions and <u>write</u> the missing numbers or words.

This is a _____.

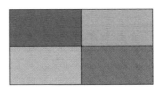

It is divided into ... parts.

$\boxed{1}$ part is $\frac{...}{...}$ (_____).

<u>How many one fourths</u> are in the rectangle?   ... .

<u>Color</u> $\frac{1}{4}$ red.   <u>Color</u> $\frac{2}{4}$ green.   <u>Color</u> $\frac{3}{4}$ blue.   <u>Color</u> $\frac{4}{4}$ yellow.

   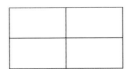

www.stemmindset.com

1. I put $\boxed{15}$ cupcakes equally on $\boxed{3}$ plates. <u>What part of the cupcakes</u> was on one plate? <u>Fill in</u> the missing numbers.

Cupcakes	Plates	Part
...	...	$? \frac{...}{...}$

$1 = \frac{...}{...}$

To solve the problem, I use a diagram. See, I have 3 plates, I make 3 line segments. I can divide 15 cupcakes equally into 3 parts: 15 = 5 + 5 + 5. Since 1 part out of three plates equals $\frac{1}{3}$, 5 cupcakes on one plate equal $\frac{...}{...}$, too.

Answer:____

2. I've read $\boxed{8}$ books equally in $\boxed{4}$ days. <u>What part of the books</u> have I read in 2 days?

Books	Days	Part
...	...	$? \frac{...}{...}$

$1 = \frac{...}{...}$

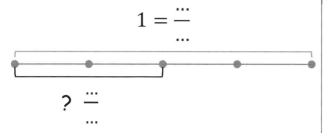

$? \frac{...}{...}$

Answer:_____

1. <u>Write</u> the missing numbers to make the fractions complete.

Let me think.

I can divide numbers 4 and 8 into 4 equal parts and 5 and 10 into 5 equal parts.

1 + 1 + 1 + 1:

$$1 \text{ whole} = \frac{4}{4} = \frac{\dots \text{ equal parts}}{\dots \text{ in the whole}}$$

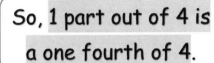

So, 1 part out of 4 is a one fourth of 4.

$$\frac{1}{4} \qquad \frac{\square}{\square} \qquad \frac{\square}{\square} \qquad \frac{\square}{\square}$$

I know that 2 parts out of 8 is one fourth of 8.

2 + 2 + 2 + 2:

$$1 \text{ whole} = \frac{8}{8} = \frac{\dots \text{ equal parts}}{\dots \text{ in the whole}}$$

$$\frac{2}{8} = \frac{1}{4} \qquad \frac{\square}{\square} = \frac{\square}{\square} \qquad \frac{\square}{\square} = \frac{\square}{\square} \qquad \frac{\square}{\square} = \frac{\square}{\square}$$

I know that 2 triangles out of 10 is one fifth out of 10.

2 + 2 + 2 + 2 + 2:

$$1 \text{ whole} = \frac{\square}{\square} = \frac{\dots \text{ equal parts}}{\dots \text{ in the whole}}$$

$$\frac{\square}{\square} = \frac{\square}{\square} \qquad \frac{\square}{\square} = \frac{\square}{\square} \qquad \frac{\square}{\square} = \frac{\square}{\square} \qquad \frac{\square}{\square} = \frac{\square}{\square} \qquad \frac{\square}{\square} = \frac{\square}{\square}$$

www.stemmindset.com

1. <u>Compare</u> the fractions using the fractions strips (">," "<," or "=.")

$\dfrac{1}{2}$ ... $\dfrac{1}{5}$

$\dfrac{1}{8}$ ... $\dfrac{2}{5}$

$\dfrac{2}{4}$ ... $\dfrac{3}{5}$

$\dfrac{1}{2}$ ... $\dfrac{2}{5}$

$\dfrac{2}{4}$ ... $\dfrac{4}{8}$

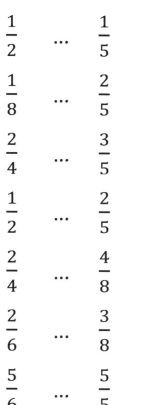

$\dfrac{2}{6}$ ... $\dfrac{3}{8}$ $\dfrac{3}{6}$ ... $\dfrac{1}{2}$ $\dfrac{4}{5}$ ... $\dfrac{4}{6}$

$\dfrac{5}{6}$ ... $\dfrac{5}{5}$ $\dfrac{1}{8}$ ... $\dfrac{1}{10}$ $\dfrac{3}{6}$ ... $\dfrac{5}{10}$

$\dfrac{3}{8}$ ... $\dfrac{4}{10}$ $\dfrac{8}{10}$ ... $\dfrac{4}{5}$ $\dfrac{2}{6}$ ... $\dfrac{1}{4}$

Can you explain me <u>why</u> $\dfrac{8}{10}$ is equal to $\dfrac{4}{5}$?

8 = 4 + 4 (twice more),

10 = 5 + 5 (twice more). Got it.

Aha… Let me think… Because the number of the parts (it's called the numerator) and the number of the whole (it's called the denominator) are exactly twice more in $\dfrac{8}{10}$ than in $\dfrac{4}{5}$.

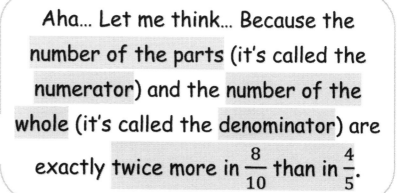

Right. They both show the same amount. $\dfrac{8}{10}$ covers the same area on the picture as $\dfrac{4}{5}$.

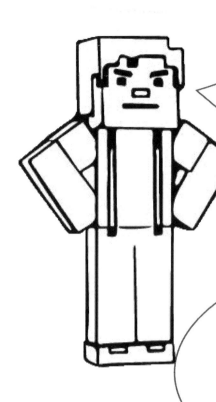

1. My denominator is even and less than 15. My numerator is odd and less than 10. I'm equivalent to one fourth. Who am I?

I know the denominator may be 2, 4, 6, 8, 10, 12, and 14. The numerator may be 1, 3, 5, 7, and 9. I'm almost there, can you help me? _____

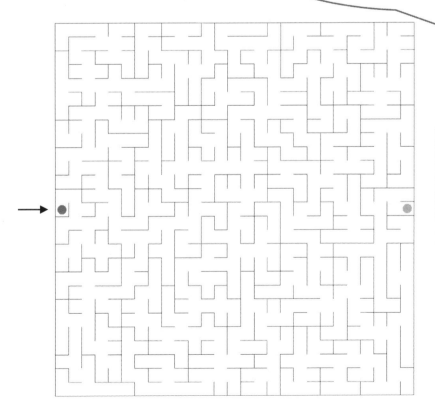

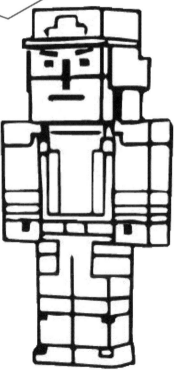

www.stemmindset.com

1.   <u>Answer</u> the questions and <u>write</u> the missing numbers.

A rectangle is divided into $\boxed{9 \text{ equal}}$ parts. Hints: 1) <u>divide</u> a rectangle into $\boxed{3 \text{ equal}}$ parts (black lines); 2) <u>divide</u> each remaining part into $\boxed{3 \text{ equal parts}}$ (blue-dotted lines). You will get $\boxed{9 \text{ equal}}$ parts.

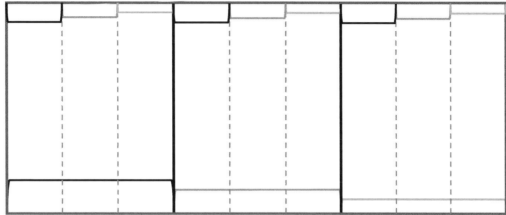

<u>What fraction names</u> $\boxed{1}$ part out of $\boxed{9}$? $1 \div 9 = \frac{1}{9}$ (one ninth).

<u>What fraction names</u> $\boxed{2}$ parts out of $\boxed{9}$? $... \div ... = \frac{...}{...}$ (_____).

<u>What fraction names</u> $\boxed{3}$ parts out of $\boxed{9}$? $... \div ... = \frac{...}{...}$ (_____).

<u>What fraction names</u> $\boxed{4}$ parts out of $\boxed{9}$? $... \div ... = \frac{...}{...}$ (_____).

<u>What fraction names</u> $\boxed{5}$ parts out of $\boxed{9}$? $... \div ... = \frac{...}{...}$ (_____).

<u>What fraction names</u> $\boxed{6}$ parts out of $\boxed{9}$? $... \div ... = \frac{...}{...}$ (_____).

<u>What fraction names</u> $\boxed{7}$ parts out of $\boxed{9}$? $... \div ... = \frac{...}{...}$ (_____).

<u>What fraction names</u> $\boxed{8}$ parts out of $\boxed{9}$? $... \div ... = \frac{...}{...}$ (_____).

<u>What fraction names</u> $\boxed{9}$ parts out of $\boxed{9}$? $... \div ... = \frac{...}{...} = 1$.

The whole always equals 1 or $\frac{2}{2}$, or $\frac{3}{3}$, or $\frac{4}{4}$, or $\frac{5}{5}$, or $\frac{9}{9}$.

1. <u>Write</u> the missing numbers and <u>subtract</u> the fractions. The first one is done for you.

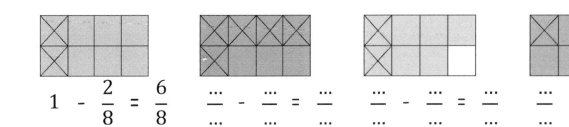

$$1 - \frac{2}{8} = \frac{6}{8}$$ ⋯ $$\frac{⋯}{⋯} - \frac{⋯}{⋯} = \frac{⋯}{⋯}$$ ⋯ $$\frac{⋯}{⋯} - \frac{⋯}{⋯} = \frac{⋯}{⋯}$$ ⋯ $$\frac{⋯}{⋯} - \frac{⋯}{⋯} = \frac{⋯}{⋯}$$

2. <u>Write</u> the missing numbers. <u>Add</u> the fractions. <u>Circle</u> ">" or "<".

$$\frac{1}{3} + \frac{1}{3} = \frac{⋯+⋯}{⋯} = \frac{⋯}{⋯}$$ > or < $$\frac{2}{3} + \frac{1}{3} = \frac{⋯+⋯}{⋯} = \frac{⋯}{⋯} = ⋯$$

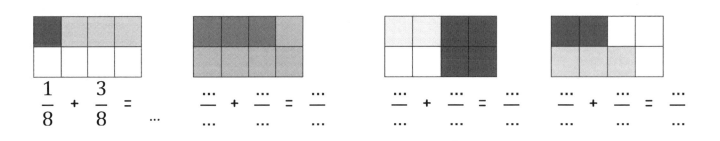

3. <u>Write</u> the missing numbers and <u>add</u> the fractions.

$$\frac{1}{8} + \frac{3}{8} = ⋯$$ $$\frac{⋯}{⋯} + \frac{⋯}{⋯} = \frac{⋯}{⋯}$$ $$\frac{⋯}{⋯} + \frac{⋯}{⋯} = \frac{⋯}{⋯}$$ $$\frac{⋯}{⋯} + \frac{⋯}{⋯} = \frac{⋯}{⋯}$$

1. <u>Subtract</u> and <u>compare</u> (">" or "<") the fractions.

$$1 - \frac{5}{6} = \frac{...}{...} - \frac{...}{...} = \frac{... - ...}{...} = \frac{...}{...}$$ ... $$1 - \frac{2}{3} = ... \frac{...}{...} - \frac{...}{...} = \frac{... - ...}{...} = \frac{...}{...}$$

2. <u>Write</u> the missing numbers and <u>subtract</u> the fractions. The first one is done for you.

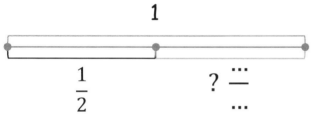

$$1 - \frac{1}{2} = \frac{2}{2} - \frac{1}{2} = \frac{2-1}{2} = \frac{...}{...}$$

$$1 - \frac{1}{4} = \frac{...}{...} - \frac{...}{...} = \frac{... - ...}{...} = \frac{...}{...}$$

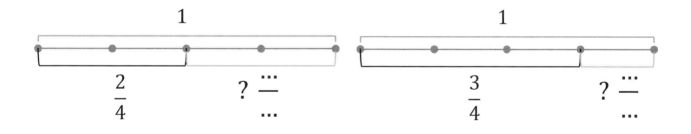

$$1 - \frac{2}{4} = \frac{...}{...} - \frac{...}{...} = \frac{... - ...}{...} = \frac{...}{...}$$

$$1 - \frac{3}{4} = \frac{...}{...} - \frac{...}{...} = \frac{... - ...}{...} = \frac{...}{...}$$

1. <u>Answer</u> the questions and <u>write</u> the missing numbers.

A rectangle is <u>divided</u> into 8 equal parts. Hints: 1) <u>divide</u> a rectangle into 2 equal parts (blue-dotted line); 2) <u>divide</u> each half into 2 equal parts (red-dotted lines); 3) <u>divide</u> each quarter into 2 equal parts (black and green-dotted lines). You will get 8 equal parts altogether.

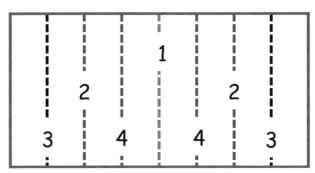

<u>What fraction names</u> 1 part out of 8? $1 \div 8 = \frac{1}{8}$ (one eighth).

<u>What fraction names</u> 2 parts out of 8? ... ÷ ... = $\frac{...}{...}$ (_____).

<u>What fraction names</u> 3 parts out of 8? ... ÷ ... = $\frac{...}{...}$ (_____).

<u>What fraction names</u> 4 parts out of 8? ... ÷ ... = $\frac{...}{...}$ (_____).

4 parts equal a half of the rectangle. So, $\frac{4}{8} = \frac{1}{2}$.

<u>What fraction names</u> 5 parts out of 8? ... ÷ ... = $\frac{...}{...}$ (_____).

<u>What fraction names</u> 6 parts out of 8? ... ÷ ... = $\frac{...}{...}$ (_____).

<u>What fraction names</u> 7 parts out of 8? ... ÷ ... = $\frac{...}{...}$ (_____).

<u>What fraction names</u> 8 parts out of 8? ... ÷ ... = $\frac{...}{...}$ = 1.

The **whole** always equals 1 or $\frac{2}{2}$, or $\frac{3}{3}$, or $\frac{4}{4}$, or $\frac{5}{5}$, $\frac{8}{8}$.

www.stemmindset.com

1. I'm less than three and a half. I'm an odd number of fourths. What is the least and the greatest number of fourths can I have?

Answer: _____

_____.

1.   <u>Add</u> and <u>compare</u> the fractions. <u>Circle</u> ">," "<," or "=."

$\dfrac{1}{2}$ >   <   = $\dfrac{3}{8}$

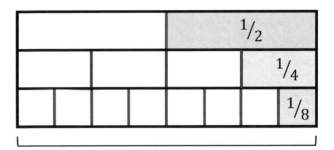

$\dfrac{2}{4}$ >   <   = $\dfrac{4}{8}$

$\dfrac{1}{2} + \dfrac{2}{2} = \dfrac{1+2}{2} = \dfrac{...}{...}$ >   <   = $\dfrac{2}{4} + \dfrac{2}{4} = \dfrac{...+...}{...} = \dfrac{...}{...}$

$\dfrac{5}{8} + \dfrac{2}{8} = \dfrac{...+...}{...} = \dfrac{...}{...}$ >   <   = $\dfrac{3}{4} + \dfrac{1}{4} = \dfrac{...+...}{...} = \dfrac{...}{...}$

2.   <u>Add</u> the fractions.

$\dfrac{1}{5} + \dfrac{1}{5} = \dfrac{...}{...}$ $\dfrac{1}{5} + \dfrac{4}{5} = \dfrac{...}{...} = ...$ $\dfrac{2}{5} + \dfrac{3}{5} = \dfrac{...}{...} = ...$

$\dfrac{3}{5} - \dfrac{1}{5} = \dfrac{...}{...}$ $\dfrac{5}{5} - \dfrac{4}{5} = \dfrac{...}{...}$ $\dfrac{4}{5} - \dfrac{1}{5} = \dfrac{...}{...}$

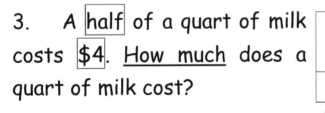

$\dfrac{1}{5}$

3.   A [half] of a quart of milk costs [$4]. <u>How much</u> does a quart of milk cost?

A half	Halves in a quart	Quart of milk
...	...	? ...

$\dfrac{1}{2} = \$...$

Answer: _____.

            www.stemmindset.com

1. <u>Answer</u> the questions and <u>write</u> the missing numbers.

A rectangle is <u>divided</u> into 4 equal parts.

Hints: 1) <u>divide</u> a rectangle into 2 equal parts (blue-dotted line); 2) <u>divide</u> each half into 2 equal parts (red-dotted lines). You will get 4 equal parts in all.

 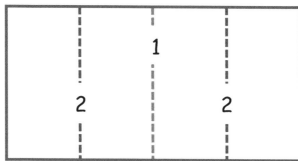

<u>What fraction names</u> 1 part out of 4 equal parts?

$$1 \div 4 = \frac{1}{4} \text{ (one fourth).}$$

<u>What fraction names</u> 2 parts out of 4?

$$\ldots \div \ldots = \frac{\ldots}{\ldots} (\underline{\hspace{5cm}}).$$

2 parts equal a half of the rectangle. So,

$$\frac{2}{4} = \frac{1}{2}.$$

<u>What fraction names</u> 3 parts out of 4? $\ldots \div \ldots = \frac{\ldots}{\ldots}.$

or 2 parts + 1 part: $\frac{2}{4} + \frac{1}{4} = \frac{\ldots}{\ldots}.$

<u>What fraction names</u> 4 parts out of 4? $\ldots \div \ldots = \frac{\ldots}{\ldots} = 1.$

The whole always equals 1 or $\frac{2}{2}$, or $\frac{3}{3}$, or $\frac{4}{4}$.

1. <u>Subtract and compare</u> the fractions (<u>use</u> ">" or "<").

$\dfrac{5}{5} - \dfrac{1}{5} = \dfrac{...}{...}$ ... $\dfrac{4}{5} - \dfrac{1}{5} = \dfrac{...}{...}$ ... $\dfrac{3}{5} - \dfrac{2}{5} = \dfrac{...}{...}$

$\dfrac{3}{5} + \dfrac{1}{5} - \dfrac{2}{5} = \dfrac{...}{...}$ ... $\dfrac{4}{5} - \dfrac{2}{5} + \dfrac{3}{5} = \dfrac{...}{...}$ ... $\dfrac{2}{5} + \dfrac{3}{5} - \dfrac{4}{5} = \dfrac{...}{...}$

$^1/_5$

2. I have 10 halves of a cup. <u>How many cups</u> do I have in the pot?

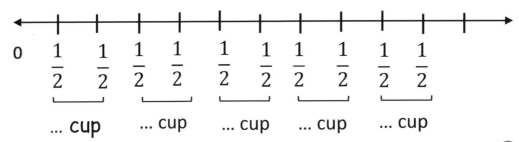

$0 \quad \dfrac{1}{2} \quad \dfrac{1}{2} \quad \dfrac{1}{2} \quad \dfrac{1}{2} \quad \dfrac{1}{2} \quad \dfrac{1}{2} \quad \dfrac{1}{2} \quad \dfrac{1}{2} \quad \dfrac{1}{2} \quad \dfrac{1}{2}$

... cup   ... cup   ... cup   ... cup   ... cup

I can add halves, or I can add cups.

1 cup = ... halves of a cup. So, 1 cup = $\dfrac{...}{...} + \dfrac{...}{...}$ =

Answer: _____.

         www.stemmindset.com

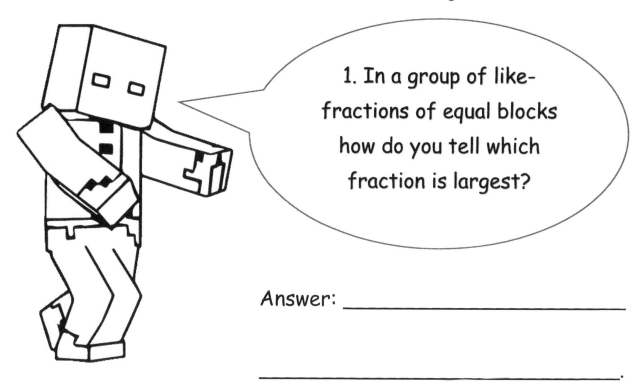

1. In a group of like-fractions of equal blocks how do you tell which fraction is largest?

Answer: _____

_____.

2. <u>Find</u> and <u>circle</u> or <u>cross out</u> the words to find out more about natural landscapes in Minecraft.

```
L  R  C  T  Z  S  N  D  R  E  A  R        BIOMES
K  L  R  O  E  C  Q  Q  F  E  S  E
W  D  A  M  L  Q  Y  N  L  G  R  H        TEMPERATURE
U  L  O  F  Z  D  R  U  A  Y  Z  T
L  I  G  T  N  I  E  D  C  I  G  A        WEATHER
B  X  J  K  V  I  A  R  P  W  C  E
B  H  U  P  J  Z  A  G  A  A  D  W        SNOWY
V  F  V  X  H  X  Q  R  W  R  K  U        COLD
N  X  J  W  Q  Q  D  F  O  M  D  Q
E  R  U  T  A  R  E  P  M  E  T  X        DRY
S  N  O  W  Y  O  H  E  O  Z  X  X        WARM
B  Z  Y  Z  O  I  K  L  X  J  M  T        RAINFALL
```

1. <u>Answer</u> the questions and <u>write</u> the missing numbers.

A rectangle is <u>divided</u> into 6 equal parts. Hints: 1) <u>divide</u> a rectangle into 2 equal parts (green-dotted line); 2) <u>divide</u> each half into 3 equal parts (red-dotted lines, then, black-dotted lines). You will get 6 equal parts in all.

 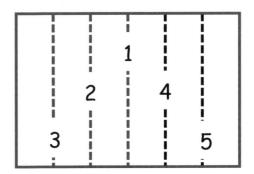

<u>What fraction names</u> 1 part out of 6? $1 \div 6 = \frac{1}{6}$ (one sixth).

<u>What fraction names</u> 2 parts out of 6? ... $\div$ ... $= \frac{...}{...}$ (_____).

<u>What fraction names</u> 3 parts out of 6? ... $\div$ ... $= \frac{...}{...}$ (_____).

3 parts equal a half of the rectangle. So, $\frac{3}{6} = \frac{1}{2}$.

<u>What fraction names</u> 4 parts out of 6? ... $\div$ ... $= \frac{...}{...}$ (_____).

<u>What fraction names</u> 5 parts out of 6? ... $\div$ ... $= \frac{...}{...}$ (_____).

The whole always equals 1 or $\frac{2}{2}$, or $\frac{3}{3}$, or $\frac{4}{4}$, or $\frac{6}{6}$.

www.stemmindset.com

1. <u>Compare</u> the fractions. <u>Use</u> "=," ">," or "<" and the fractions strips below.

$\frac{3}{4}$ ... $\frac{3}{8}$ $\frac{2}{8}$ ... $\frac{2}{12}$

$\frac{1}{4}$ ... $\frac{2}{8}$ $\frac{2}{4}$ ... $\frac{4}{12}$

$\frac{9}{12}$ ... $\frac{3}{4}$ $\frac{2}{8}$ ... $\frac{3}{12}$

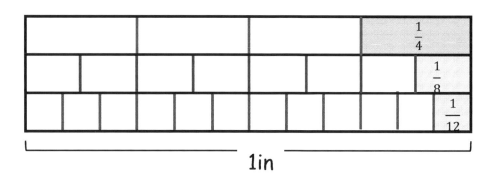

1in

2. <u>Answer</u> the questions.

<u>How many parts</u> are in this bar?... parts.

<u>What fraction names</u> $\boxed{1}$ block? $\frac{...}{...}$

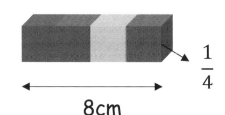

$\frac{1}{4}$

8cm

<u>How long</u> is $\frac{1}{4}$ if the bar is 8 cm long?   ... cm.

<u>How long</u> is $\frac{2}{4}$ if the bar is 8 cm long? _____ ... cm.

<u>How long</u> is $\frac{3}{4}$ if the bar is 8 cm long? _____ ... cm.

1.   <u>Answer</u> the questions.

<u>How many parts</u> are in this bar?

... parts.

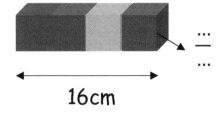

<u>What fraction names</u> 1 block? $\frac{...}{...}$.

16cm

<u>How long</u> is $\frac{1}{4}$ if the bar is 16 cm long?   ... cm.

<u>How long</u> is $\frac{2}{4}$ if the bar is 16 cm long? _____ ... cm.

<u>How long</u> is $\frac{3}{4}$ if the bar is 16 cm long? _____ ... cm.

2.   <u>How many more</u> do you need to add to make 1 whole ? <u>Write</u> the missing fraction.

$\frac{1}{2} + \frac{...}{...}$ $\frac{3}{4} + \frac{...}{...}$ $\frac{5}{6} + \frac{...}{...}$ $\frac{2}{3} + \frac{...}{...}$ $\frac{9}{10} + \frac{...}{...}$ $\frac{4}{5} + \frac{...}{...}$ $\frac{7}{8} + \frac{...}{...}$

3.   <u>Write</u> the missing fractions.

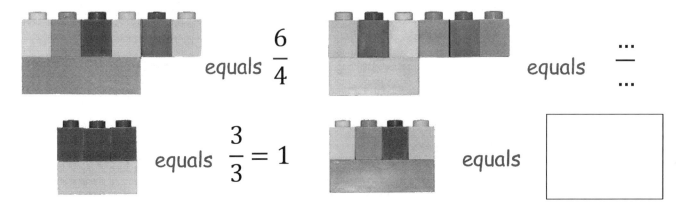

equals $\frac{6}{4}$ equals $\frac{...}{...}$

equals $\frac{3}{3} = 1$ equals

             www.stemmindset.com

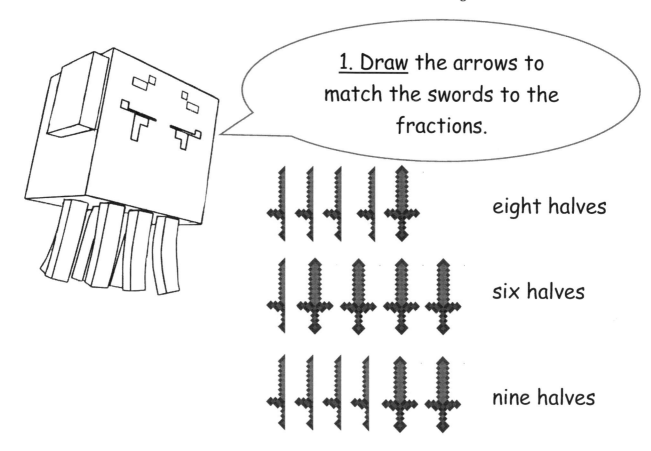

1. Draw the arrows to match the swords to the fractions.

eight halves

six halves

nine halves

2. Answer the question.

A	B	C	D	E	F	G	H	I	J	K	L	M
1	...	...	...	...	...	...	...	...	...	...	...	...
N	O	P	Q	R	S	T	U	V	W	X	Y	Z
...	...	...	...	...	...	...	...	...	...	...	...	26

‾23 ‾8 ‾1 ‾20   ‾4 ‾15   ‾25 ‾15 ‾21   ‾14 ‾5 ‾5 ‾4   ‾20 ‾15

‾3 ‾18 ‾1 ‾6 ‾20   ‾1   ‾20 ‾15 ‾18 ‾3 ‾8_?

_____.

1. <u>Answer</u> the questions and <u>write</u> the missing numbers.

<u>How many parts</u> are in this bar?   ... parts.

<u>What fraction names</u> $\boxed{1}$ block? $\dfrac{...}{...}$.

<u>How long</u> is $\dfrac{1}{8}$ if the bar 80 cm long?   ... cm.

<u>How</u> did you find it?   _____.

<u>How long</u> is $\dfrac{3}{8}$?   ... cm.

<u>How long</u> is $\dfrac{6}{8}$?   ... cm.

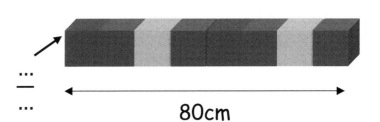

$\dfrac{...}{...}$
...   80cm

2. <u>Add</u> the fractions.

$\dfrac{2}{3} + \dfrac{3}{3} = \dfrac{...}{...}$ $\dfrac{1}{3} + \dfrac{2}{3} = \dfrac{...}{...}$ $\dfrac{2}{3} + \dfrac{2}{3} = \dfrac{...}{...}$

$\dfrac{1}{3} + \dfrac{1}{3} = \dfrac{...}{...}$ $\dfrac{1}{3} + \dfrac{4}{3} = \dfrac{...}{...}$ $\dfrac{3}{3} + \dfrac{3}{3} = \dfrac{...}{...}$

$1 + \dfrac{3}{3} = \dfrac{...}{...}$ $3 + \dfrac{1}{3} = \dfrac{...}{...}$ $2 + \dfrac{2}{3} = \dfrac{...}{...}$

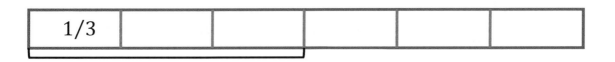

1/3					

                   www.stemmindset.com

1.   <u>Write</u> the missing numbers.

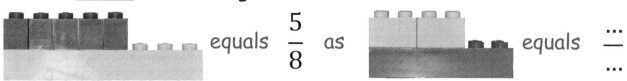

equals $\dfrac{5}{8}$ as

equals $\dfrac{...}{...}$

equals $\dfrac{3}{4}$ as

equals $\dfrac{...}{...}$

2.   <u>Answer</u> the questions.

<u>Sketch</u> a circle in the box and <u>divide</u> it into ⎡5 equal⎤ parts. <u>Compare</u> the fractions using "<," ">," or "=:"

$\dfrac{1}{5}$ ... $\dfrac{4}{5}$?

$\dfrac{3}{5}$ ... $\dfrac{2}{5}$?

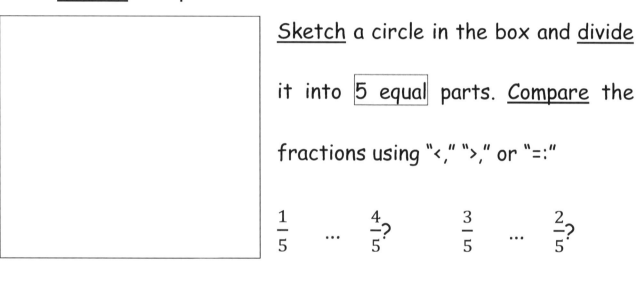

3.   <u>Compare</u> the fractions using ">," "<," or "=."

$\dfrac{5}{6}$ ... $\dfrac{2}{3}$

$\dfrac{4}{6}$ ... $\dfrac{2}{3}$

$\dfrac{3}{3}$ ... $\dfrac{6}{6}$

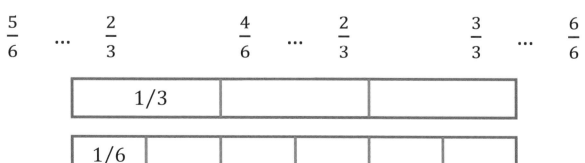

	1/3		

1/6					

1.   <u>Evaluate</u> each equation.

$1 - \dfrac{2}{3} = \dfrac{...}{...}$
$\qquad\qquad$
$1 - \dfrac{1}{3} = \dfrac{...}{...}$
$\qquad\qquad$
$1 - \dfrac{3}{3} = \dfrac{...}{...} = ...$

$\dfrac{2}{3} - \dfrac{1}{3} = \dfrac{...}{...}$
$\qquad\qquad$
$\dfrac{4}{3} - \dfrac{1}{3} = \dfrac{...}{...} = ...$
$\qquad\qquad$
$\dfrac{3}{3} - \dfrac{2}{3} = \dfrac{...}{...}$

$\dfrac{3}{3} + \dfrac{1}{3} - \dfrac{2}{3} = \dfrac{...}{...}$
$\qquad$
$\dfrac{4}{3} - \dfrac{2}{3} - \dfrac{1}{3} = \dfrac{...}{...}$
$\qquad$
$\dfrac{2}{3} + \dfrac{3}{3} - \dfrac{4}{3} = \dfrac{...}{...}$

2.   <u>Take</u> a ruler, <u>measure</u> the rectangle, and <u>answer</u> the questions.

<u>Divide</u> a rectangle vertically into $\boxed{6}$ equal parts.

<u>What fraction names</u> $\boxed{1}$ part of the rectangle?
$\qquad\qquad\qquad\qquad$ $\dfrac{...}{...}$.

<u>What kinds of shapes</u> did you get?

_____.

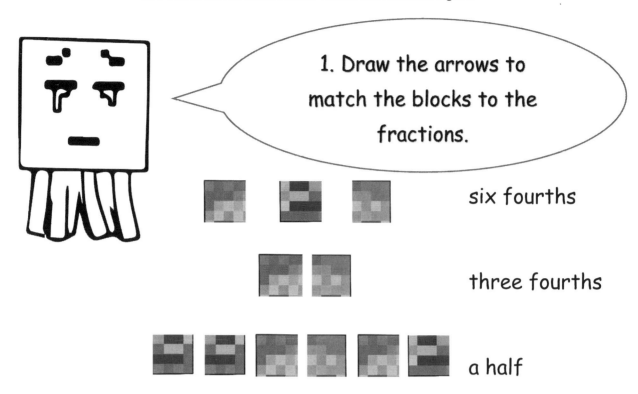

1. Draw the arrows to match the blocks to the fractions.

six fourths

three fourths

a half

1.   <u>Compare</u> the fractions using ">," "<," or "=."

$\frac{1}{2}$ ... $\frac{4}{6}$ $\frac{1}{6}$ ... $\frac{1}{3}$ $\frac{2}{3}$ ... $\frac{4}{6}$

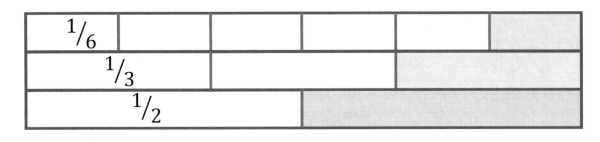

2.   <u>Add or subtract</u> the fractions.

$\frac{1}{6} + \frac{2}{6} =$ $\frac{1}{6} + \frac{3}{6} =$ $\frac{4}{6} + \frac{1}{6} =$

$1 + \frac{5}{6} =$ $3 + \frac{1}{6} =$ $2 + \frac{2}{6} =$

$1 - \frac{2}{6} =$ $1 - \frac{1}{6} =$ $1 - \frac{3}{6} =$

$\frac{1}{6}$

3.   <u>Add</u> the fractions.

$\frac{2}{5} + \frac{3}{5} =$ $\frac{1}{5} + \frac{2}{5} =$ $\frac{2}{5} + \frac{2}{5} =$

$1 + \frac{2}{5} =$ $5 + \frac{5}{5} =$ $1 + \frac{1}{5} =$

                       www.stemmindset.com

1. <u>Add and subtract</u> the fractions.

$\frac{2}{6} - \frac{1}{6} =$ $\qquad$ $\frac{4}{6} - \frac{1}{6} =$ $\qquad$ $\frac{2}{6} + \frac{3}{6} - \frac{4}{6} =$

$\frac{3}{6} + \frac{1}{6} - \frac{2}{6} =$ $\qquad$ $\frac{4}{6} - \frac{2}{6} - \frac{1}{6} =$ $\qquad$ $\frac{3}{6} - \frac{2}{6} =$

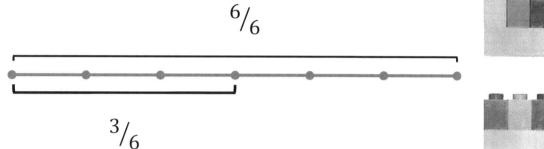

2. <u>Compare</u> the fractions using the fractions strips below, <u>fill in</u>

">," "<," or "=."

$\frac{1}{3}$ ... $\frac{2}{6}$ $\qquad\qquad\qquad$ $\frac{4}{6}$ ... $\frac{3}{3}$

$\frac{1}{3}$ ... $\frac{1}{2}$ $\qquad\qquad\qquad$ $\frac{5}{6}$ ... $\frac{3}{6}$

1/6					
1/3		1/3			
1/2					

1.   Write the missing fraction.

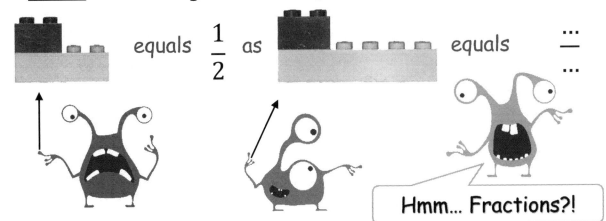

equals $\frac{1}{2}$ as   equals $\frac{...}{...}$

Hmm... Fractions?!

2.   Add and subtract the fractions.

$\frac{4}{6} + \frac{2}{6} =$ $\frac{2}{6} + \frac{3}{6} =$ $\frac{1}{6} + \frac{1}{6} =$

$1 + \frac{3}{6} =$ $3 + \frac{6}{6} =$ $2 + \frac{4}{6} =$

$2 - \frac{6}{6} =$ $1 - \frac{4}{6} =$ $1 - \frac{5}{6} =$

3.   Evaluate each equation.

$\frac{5}{5} - \frac{1}{5} =$ $\frac{4}{5} - \frac{1}{5} =$ $\frac{3}{5} - \frac{2}{5} =$

$\frac{3}{5} + \frac{1}{5} - \frac{2}{5} =$ $\frac{4}{5} - \frac{2}{5} + \frac{3}{5} =$ $\frac{2}{5} + \frac{3}{5} - \frac{4}{5} =$

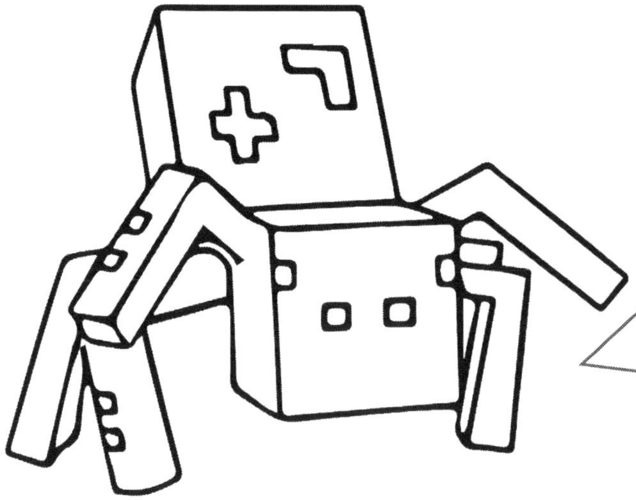

1. I met 3 wolves, 5 pigs, 6 cows, and 4 dogs. What fraction of the animals were pigs?

Answer:

_____.

2. <u>Unscramble</u> each word to find out more about biomes in Minecraft. <u>Write</u> the word in the squares.

retarin

ecahb

iatga

uaimtonns

nalsmawpd

1.   <u>Answer</u> the questions and <u>write</u> the missing numbers.

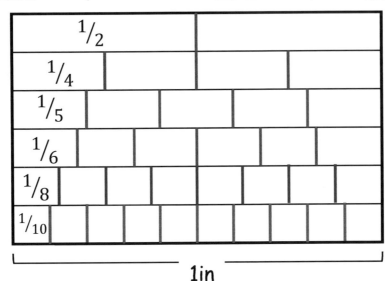

1in

<u>How many equal parts</u> are in 1 inch (the whole)?

$\frac{1}{2}$?   2 $\frac{1}{6}$?   ...

$\frac{1}{4}$?   ... $\frac{1}{8}$?   ...

$\frac{1}{5}$?   ... $\frac{1}{10}$?   ...

<u>How many equal parts</u> are in a half of the whole (or an inch)?

$\frac{1}{4}$?   ... $\frac{1}{6}$?   ... $\frac{1}{8}$?   ... $\frac{1}{10}$?   ...

<u>Write</u> the missing numerators: $1 = \frac{...}{2} = \frac{...}{10} = \frac{...}{3} = \frac{...}{6} = \frac{...}{5} = \frac{...}{8} = \frac{...}{4}$

<u>How many</u> $\frac{1}{10}$'s are in $\frac{3}{5}$ of an inch?   ...

<u>How many</u> $\frac{1}{8}$'s are in $\frac{3}{4}$ of an inch?   ...

                               www.stemmindset.com

1. <u>Answer</u> the questions.

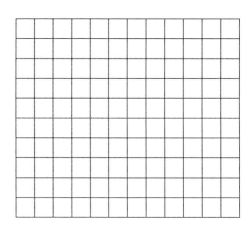

<u>Draw</u> a rectangle in the box and <u>divide</u> it into $\boxed{5}$ equal parts. <u>Compare</u> the fractions using "<," ">," or "=:"

$\dfrac{5}{5}$ ... 1? $\dfrac{2}{5}$ ... $\dfrac{4}{5}$?

2. <u>What fraction name</u> do you need to subtract to get $\boxed{1\text{ whole}}$?

$\dfrac{2}{2} = \dfrac{3}{3} = \dfrac{4}{4} = \dfrac{5}{5} = \dfrac{6}{6} = \dfrac{8}{8} = \dfrac{10}{10} = 1.$

$\dfrac{14}{10} - \dfrac{4}{10}$ $\dfrac{7}{4} - \dfrac{...}{...}$ $\dfrac{9}{6} - \dfrac{...}{...}$ $\dfrac{5}{3} - \dfrac{...}{...}$

$\dfrac{3}{2} - \dfrac{...}{...}$ $\dfrac{8}{5} - \dfrac{...}{...}$ $\dfrac{10}{8} - \dfrac{...}{...}$

3. <u>Add and subtract</u> the fractions.

$\dfrac{5}{6} - \dfrac{1}{6} =$ $\dfrac{4}{6} - \dfrac{3}{6} =$ $\dfrac{5}{6} - \dfrac{3}{6} =$

$\dfrac{5}{6} + \dfrac{1}{6} - \dfrac{3}{6} =$ $\dfrac{5}{6} - \dfrac{1}{6} - \dfrac{2}{6} =$ $\dfrac{1}{6} + \dfrac{5}{6} - \dfrac{2}{6} =$

1.　Let's make an experiment! You need 3 cups, a jar with water, and an empty jar. Fill in the missing fractions or numbers.

1) Fill 1 cup with water. You have a WHOLE cup of water or it equals ….

2) Take another cup and fill a half of it with water. You have $\frac{...}{...}$ of a cup of water.

3) Pour the first and second cups into the empty jar. You poured … cup and $\frac{...}{...}$ of a cup which equals …$\frac{...}{...}$.

　… cup　　　　$\frac{...}{...}$ of a cup　　　　　　…$\frac{...}{...}$ of a cup

Compare the 2 cups and the jar and use ">," "<," or "=."

　　　　　　…　　　　　　…

Compare the fractions and fill in ">" or "<":

1　…　$\frac{1}{2}$;　　$\frac{1}{2}$　…　$1\frac{1}{2}$;　　1　…　$1\frac{1}{2}$.

　　www.stemmindset.com

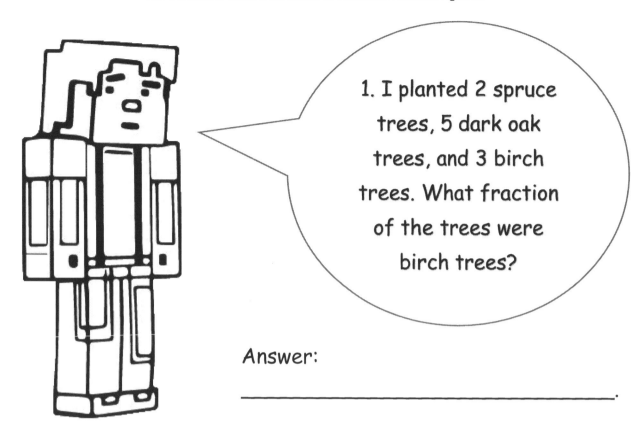

1. I planted 2 spruce trees, 5 dark oak trees, and 3 birch trees. What fraction of the trees were birch trees?

Answer:

_____.

2. <u>Find</u> and <u>circle</u> or <u>cross out</u> the words to find out more about blazes in Minecraft.

```
R  M  R  M  O  B  B  Q  L  F  D  C     GRAY
Q  E  I  E  D  J  A  T  S  I  I  N
T  L  S  A  N  Q  E  T  V  R  B  B     HOSTILE
J  J  N  I  Z  W  A  N  Y  E  Y  L
V  P  M  F  S  I  A  J  N  B  B  S     MOB
P  E  P  D  R  T  O  P  S  A  S  O
Y  W  U  C  H  F  A  W  S  L  S  T     FIREBALLS
B  H  A  Y  C  W  P  N  L  L  J  A     CATCH
Z  S  S  Y  T  N  Y  I  C  S  N  R
E  B  A  J  A  P  C  A  D  E  O  H     RESISTANCE
J  R  C  Y  C  E  L  I  T  S  O  H     STAIRCASE
G  C  J  O  L  C  D  E  Z  I  C  V     SPAWNER
```

1. <u>Write</u> the missing numbers and <u>subtract</u> the fractions.

1/6				

$\dfrac{5}{6} - \dfrac{2}{6} = \dfrac{... - ...}{...} = \dfrac{...}{...}$ $\dfrac{4}{6} - \dfrac{3}{6} = \dfrac{... - ...}{...} = \dfrac{...}{...}$

1/3		

$\dfrac{2}{3} - \dfrac{1}{3} = \dfrac{... - ...}{...} = \dfrac{...}{...}$ $\dfrac{3}{3} - \dfrac{1}{3} = \dfrac{... - ...}{...} = \dfrac{...}{...}$

2. <u>Compare</u> the fractions using ">," "<," or "=."

1/3		

1/6				

$\dfrac{2}{6}$ ... $\dfrac{1}{3}$ $\dfrac{5}{6}$ ... $\dfrac{2}{3}$

$\dfrac{1}{6}$ ... $\dfrac{1}{3}$ $\dfrac{2}{6}$ ... $\dfrac{2}{3}$

$\dfrac{4}{6}$ ... $\dfrac{2}{3}$ $\dfrac{3}{3}$ ... $\dfrac{3}{6}$

     www.stemmindset.com

1. <u>Add and compare</u> the fractions using ">," "<," or "=."

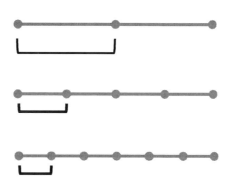

$$\frac{1}{2} + \frac{1}{2} = \frac{...+...}{...} = \frac{...}{...}$$ ... $$\frac{1}{4} + \frac{1}{4} = \frac{...+...}{...} = \frac{...}{...}$$

$$\frac{1}{6} + \frac{1}{6} = \frac{...+...}{...} = \frac{...}{...}$$ ... $$\frac{2}{6} + \frac{1}{6} = \frac{...+...}{...} = \frac{...}{...}$$

$$\frac{3}{6} + \frac{3}{6} = \frac{...+...}{...} = \frac{...}{...}$$ ... $$\frac{1}{4} + \frac{3}{4} = \frac{...+...}{...} = \frac{...}{...}$$

2. <u>Compare</u> the fractions using ">," "<," or "=."

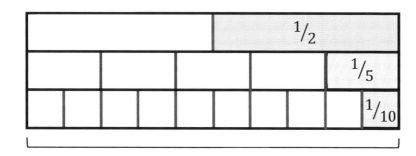

$$\frac{1}{2} \quad ... \quad \frac{5}{10}$$

$$\frac{3}{5} \quad ... \quad \frac{6}{10}$$

$$\frac{1}{2} \quad ... \quad \frac{2}{5} \qquad \frac{4}{5} \quad ... \quad \frac{7}{10} \qquad \frac{5}{5} \quad ... \quad \frac{9}{10}$$

1. <u>Fill in</u> the missing numbers and <u>draw</u> the missing shapes in the box.

 equals $1\frac{3}{4}$ as  equals $...\frac{...}{...}$

 equals $2\frac{1}{2}$ as  equals $...\frac{...}{...}$

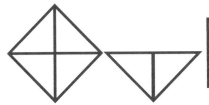 equals $2\frac{2}{4}$

as [   ] equals $4\frac{1}{4}$

2. <u>Compare</u> the fractions.

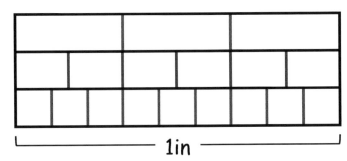

1in

$\frac{3}{3}$ ... $\frac{4}{6}$ $\frac{2}{9}$ ... $\frac{2}{6}$

$\frac{5}{6}$ ... $\frac{5}{9}$ $\frac{6}{9}$ ... $\frac{4}{6}$

1. I crafted an axe with ⬜3⬜ oak blocks and ⬜2⬜ sticks. <u>What fraction of an axe is made from sticks?</u>

Answer:_____

_____.

1. <u>Read</u> and <u>write</u> the missing numbers.

Adding like fractions (fractions with like denominators:

rewrite the denominator and add the numerators:

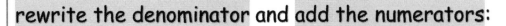

$$\frac{1}{3} + \frac{1}{3} = \frac{1+1}{3} = \frac{...}{3}$$

I add like fractions, and the numerator equals the denominator (3 = 3). That means I have 1 whole:

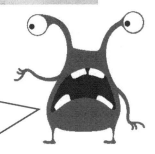

$$\frac{1}{3} + \frac{2}{3} = \frac{... + ...}{3} = \frac{...}{3} = ...$$

I add like fractions, and the numerator is more than the denominator (5 > 3). I have an improper fraction:

$$\frac{2}{3} + \frac{3}{3} = \frac{... + ...}{3} = \frac{...}{3} = ...\frac{...}{3}$$

1. <u>Write</u> the missing numbers.

I add a mixed number (the sum of a whole number and a fraction) and a like fraction. First, I add the fractions and then, I add the wholes. Last, I add the answers:

$2\frac{1}{3} + \frac{1}{3} = ...\frac{...}{3}$

1) $\frac{1}{3} + \frac{1}{3} = \frac{... + ...}{3} = \frac{...}{3}$

2) $2 + 0 = ...$

I added the fractions and got an improper fraction (three thirds), that equals 1 whole. I added the wholes (2 + 0) and got 2. Last, I added the answers (1 + 2):

$2\frac{1}{3} + \frac{2}{3} = ...$

1) $\frac{1}{3} + \frac{2}{3} = \frac{... + ...}{3} = \frac{...}{3} = ...$

2) $2 + 0 = ...$

1. <u>Write</u> the missing numbers.

When you subtract like fractions, you rewrite the denominators and subtract the numerators:

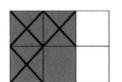

$$\frac{4}{6} - \frac{3}{6} = \frac{4-3}{6} = \frac{...}{6}$$

When you subtract a fraction out of a whole number, change a whole number to a fraction, rewrite the denominator and subtract the numerators:

$$1 - \frac{5}{6} = \frac{6}{6} - \frac{5}{6} = \frac{...-...}{6} = \frac{...}{6}$$

2. <u>Compare</u> the fractions using "=," ">," or "<."

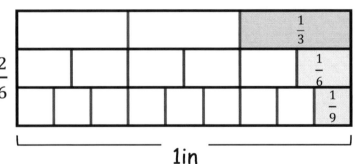

1in

$$\frac{3}{6} \quad ... \quad \frac{3}{9} \qquad\qquad \frac{2}{3} \quad ... \quad \frac{2}{6}$$

$$\frac{6}{6} \quad ... \quad \frac{3}{3} \qquad\qquad\qquad\qquad \frac{4}{6} \quad ... \quad \frac{2}{3}$$

$$\frac{4}{9} \quad ... \quad \frac{3}{6} \qquad\qquad\qquad\qquad \frac{3}{3} \quad ... \quad \frac{9}{9}$$

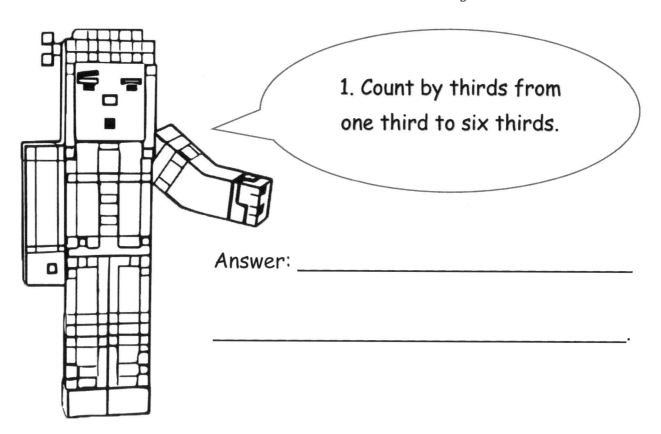

1. Count by thirds from one third to six thirds.

Answer: _____

_____.

2. <u>Unscramble</u> each word to find out more about minecarts in Minecraft. <u>Write</u> the word in the squares.

rnfceau

grkat

tacaht

nroclot

atitcarov

1. Find out how long the line segments are. $\boxed{1 \text{ square} = \frac{1}{2} \text{ of a foot}}$.

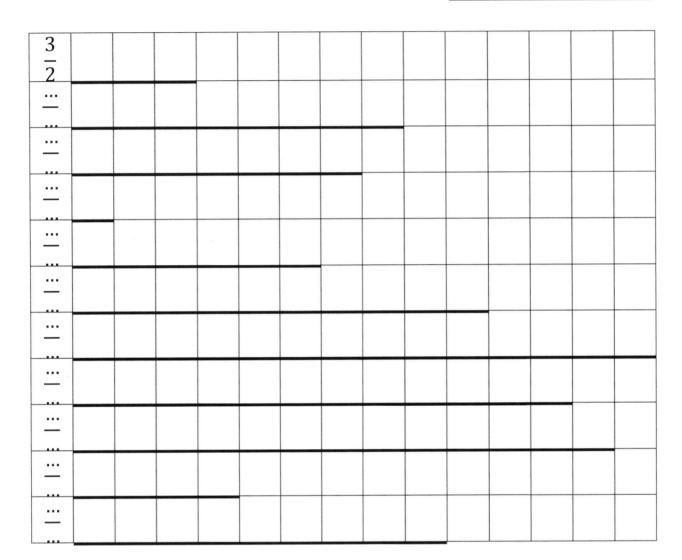

2. Draw $\boxed{2}$ lines to cut each leaf into $\boxed{4}$ equal parts ($\boxed{2}$ different ways).

www.stemmindset.com

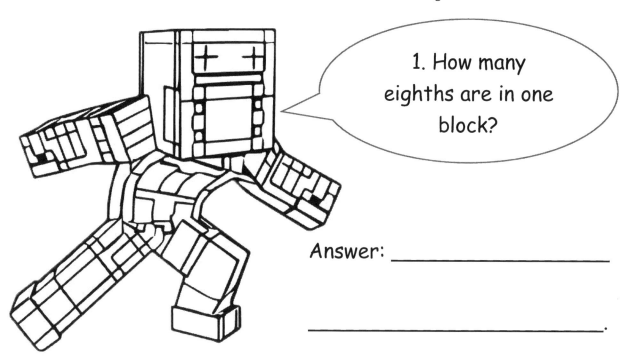

1. How many eighths are in one block?

Answer: _____

_____.

2. <u>Answer</u> the tricky question.

A	B	C	D	E	F	G	H	I	J	K	L	M
1	…	…	…	…	…	…	…	…	…	…	…	…
N	O	P	Q	R	S	T	U	V	W	X	Y	Z
…	…	…	…	…	…	…	…	…	…	…	…	26

‾ ‾ ‾ ‾   ‾ ‾   ‾ ‾ ‾   ‾ ‾ ‾ ‾   ‾ ‾
23 8 1 20   4 15   25 15 21   14 5 5 4   20 15

‾ ‾ ‾ ‾ ‾   ‾   ‾ ‾ ‾ ‾ ‾   ‾ ‾ ‾ ‾ ‾?
3 18 1 6 20   1   8 15 18 19 5   1 18 13 15 18

_____.

1.   <u>What fraction names</u> the dark part in each shape below? <u>Watch out for the tricks!</u>

Another workbook gone! We're so proud of your hard work!

www.stemmindset.com

1. <u>Read and write</u> the missing numbers.

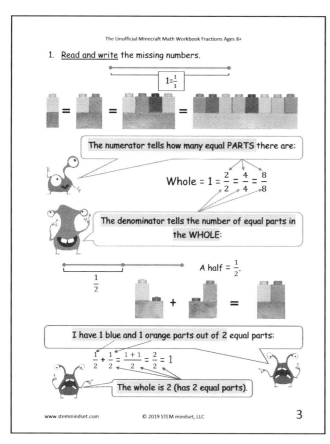

$1 = \frac{1}{1}$

The numerator tells how many equal PARTS there are:

$$\text{Whole} = 1 = \frac{2}{2} = \frac{4}{4} = \frac{8}{8}$$

The denominator tells the number of equal parts in the WHOLE:

A half = $\frac{1}{2}$.

$\frac{1}{2}$

I have 1 blue and 1 orange parts out of 2 equal parts:

$$\frac{1}{2} + \frac{1}{2} = \frac{1 + 1}{2} = \frac{2}{2} = 1$$

The whole is 2 (has 2 equal parts).

---

1. <u>Answer</u> the questions and <u>write</u> the missing numbers.

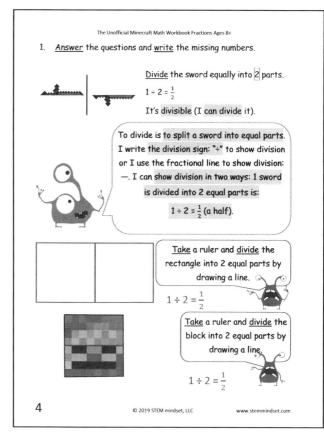

<u>Divide</u> the sword equally into 2 parts.

$1 \div 2 = \frac{1}{2}$

It's divisible (I can divide it).

To divide is to split a sword into equal parts. I write the division sign: "÷" to show division or I use the fractional line to show division: —. I can show division in two ways: 1 sword is divided into 2 equal parts is:

$1 \div 2 = \frac{1}{2}$ (a half).

<u>Take</u> a ruler and <u>divide</u> the rectangle into 2 equal parts by drawing a line.

$1 \div 2 = \frac{1}{2}$

<u>Take</u> a ruler and <u>divide</u> the block into 2 equal parts by drawing a line.

$1 \div 2 = \frac{1}{2}$

---

1. <u>Answer</u> the questions and <u>fill in</u> the missing numbers.

I have 2 halves ($\frac{1}{2}$) of a sword.

<u>How many</u> do I have in all?

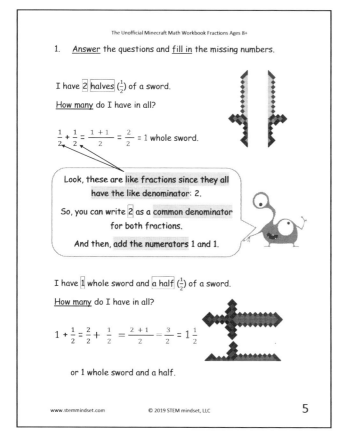

$$\frac{1}{2} + \frac{1}{2} = \frac{1 + 1}{2} = \frac{2}{2} = 1 \text{ whole sword.}$$

Look, these are like fractions since they all have the like denominator: 2.

So, you can write 2 as a common denominator for both fractions.

And then, add the numerators 1 and 1.

I have 1 whole sword and a half ($\frac{1}{2}$) of a sword.

<u>How many</u> do I have in all?

$$1 + \frac{1}{2} = \frac{2}{2} + \frac{1}{2} = \frac{2 + 1}{2} = \frac{3}{2} = 1\frac{1}{2}$$

or 1 whole sword and a half.

Great idea! I split my sword into 2 equal parts!

Are you kidding me? They are not equal! These are unequal pieces! Maybe you could try a little harder in math.

1. <u>Answer</u> the questions and <u>write</u> the missing numbers.

I have 1 whole sword and a half ($\frac{1}{2}$) of a sword. If I add a half ($\frac{1}{2}$) of a sword <u>how many</u> do I have in all? (You can choose which algorithm you like the most).

$$1\frac{1}{2} + \frac{1}{2} = \frac{3}{2} + \frac{1}{2} = \frac{3+1}{2} = \frac{4}{2} = 2 \text{ swords}$$

Add fractions   or   Add wholes

$$1\frac{1}{2} + \frac{1}{2} = 2 \qquad 1) \frac{1}{2} + \frac{1}{2} = \frac{1+1}{2} = \frac{2}{2} = 1 \qquad 2) 1 + 0 = 1$$

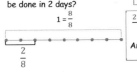

2. On Monday I get my homework for the week, usually I do $\frac{2}{8}$ of my homework per day. <u>How much of my homework</u> will be done in 2 days?

Per day	Days	Homework
$\frac{2}{8}$	2	? $\frac{4}{8}$

$$1 = \frac{8}{8}$$

$\frac{2}{8}$

$$\frac{2+2}{8} = \frac{4}{8}$$

Answer: $\frac{4}{8}$.

---

1. <u>Answer</u> the questions and <u>write</u> the missing numbers and words.

I had a block. I shared a half with my friend. <u>What part</u> is left?

$$1 - \frac{1}{2} = \frac{2}{2} - \frac{1}{2} = \frac{2-1}{2} = \frac{1}{2} \text{ (}\underline{a \text{ half of a block}}\text{)}.$$

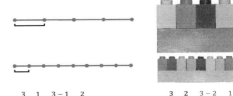

2. <u>Subtract</u> the fractions. The first one is done for you.

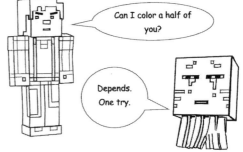

$$\frac{3}{4} - \frac{1}{4} = \frac{3-1}{4} = \frac{2}{4} \qquad\qquad \frac{3}{4} - \frac{2}{4} = \frac{3-2}{4} = \frac{1}{4}$$

$$\frac{6}{8} - \frac{3}{8} = \frac{6-3}{8} = \frac{3}{8} \qquad\qquad \frac{3}{8} - \frac{1}{8} = \frac{3-1}{8} = \frac{2}{8}$$

$$\frac{7}{8} - \frac{5}{8} = \frac{7-5}{8} = \frac{2}{8} \qquad\qquad \frac{5}{8} - \frac{1}{8} = \frac{5-1}{8} = \frac{4}{8}$$

---

1. <u>Answer</u> the questions. <u>Fill in</u> the missing numbers and words.

I had 1 whole block and a half ($\frac{1}{2}$) of a block.

I shared $\frac{1}{2}$ of a block with a friend. <u>What part</u> is left?

You can <u>choose</u> one of the two strategies:

$$1\frac{1}{2} - \frac{1}{2} = \frac{3}{2} - \frac{1}{2} = \frac{3-1}{2} = \frac{2}{2} = \text{ or } 1 \text{ whole block}\underline{\hspace{2cm}}.$$

or

$$1\frac{1}{2} - \frac{1}{2} = 1 \qquad 1) \frac{1}{2} - \frac{1}{2} = \frac{1-1}{2} = 0 \qquad 2) 1 - 0 = 1$$

2. <u>Write</u> the missing numbers. <u>Add and compare</u> the fractions using ">" or "."

$$\frac{1}{-} + \frac{1}{-} = \frac{1+1}{-} = \frac{2}{-} \qquad < \qquad \frac{3}{-} + \frac{2}{-} = \frac{3+2}{-} = \frac{5}{-}$$

Can I color a half of you?

Depends. One try.

1. <u>Find</u> and <u>circle</u> or <u>cross out</u> the words to find out more about Minecraft.

U	M	B	C	S	K	E	C	Y	M	E	H
L	N	J	E	N	P	O	E	E	I	H	F
B	I	D	A	W	R	P	J	C	N	T	U
S	X	L	E	R	B	S	M	N	E	K	W
W	P	F	I	R	T	O	J	E	S	W	Z
I	C	D	S	O	G	O	C	F	H	W	S
Z	O	F	L	O	O	R	Q	Q	A	C	Y
R	H	K	V	T	B	W	Q	H	F	Y	U
G	N	I	L	I	E	C	X	U	T	X	F
E	N	S	C	Z	Z	Z	N	Z	N	E	C
H	O	H	P	Q	J	R	K	M	G	D	B
O	O	N	T	Z	A	K	O	L	D	R	K

CORRIDOR

COBWEB

MINESHAFT

FENCE

UNDERGROUND

PLANK

CEILING

FLOOR

1. <u>Read and write</u> the missing numbers.

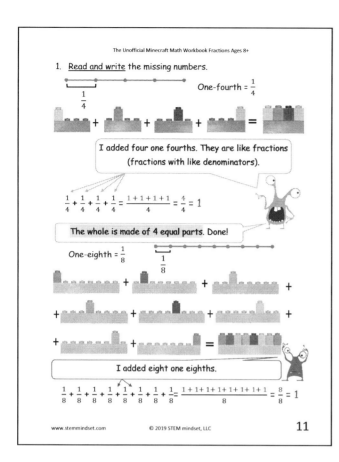

One-fourth = $\frac{1}{4}$

$\frac{1}{4}$

I added four one fourths. They are like fractions (fractions with like denominators).

$\frac{1}{4} + \frac{1}{4} + \frac{1}{4} + \frac{1}{4} = \frac{1+1+1+1}{4} = \frac{4}{4} = 1$

The whole is made of 4 equal parts. Done!

One-eighth = $\frac{1}{8}$

$\frac{1}{8}$

I added eight one eighths.

$\frac{1}{8} + \frac{1}{8} + \frac{1}{8} + \frac{1}{8} + \frac{1}{8} + \frac{1}{8} + \frac{1}{8} + \frac{1}{8} = \frac{1+1+1+1+1+1+1+1}{8} = \frac{8}{8} = 1$

---

1. <u>Fill in</u> the missing numbers.

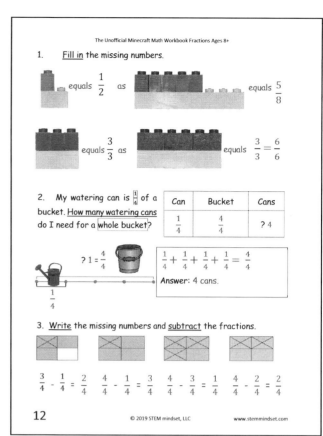

equals $\frac{1}{2}$ as   equals $\frac{5}{8}$

equals $\frac{3}{3}$ as   equals $\frac{3}{3} = \frac{6}{6}$

2. My watering can is $\frac{1}{4}$ of a bucket. <u>How many watering cans</u> do I need for a <u>whole bucket?</u>

Can	Bucket	Cans
$\frac{1}{4}$	$\frac{4}{4}$	? 4

? 1 = $\frac{4}{4}$

$\frac{1}{4}$

$\frac{1}{4} + \frac{1}{4} + \frac{1}{4} + \frac{1}{4} = \frac{4}{4}$

Answer: 4 cans.

3. <u>Write</u> the missing numbers and <u>subtract</u> the fractions.

$\frac{3}{4} - \frac{1}{4} = \frac{2}{4}$ $\frac{4}{4} - \frac{1}{4} = \frac{3}{4}$ $\frac{4}{4} - \frac{3}{4} = \frac{1}{4}$ $\frac{4}{4} - \frac{2}{4} = \frac{2}{4}$

---

1. <u>Compare</u> the fractions using ">," "<," or "=."

$\frac{1}{3} < \frac{4}{3}$ $\frac{3}{3} > \frac{2}{3}$ $\frac{2}{3} > \frac{1}{3}$

$\frac{4}{3} > \frac{2}{3}$ $\frac{3}{3} = 1$ $\frac{2}{2} = \frac{3}{3}$

$\frac{1}{3}$ $\frac{1}{2}$

2. <u>Answer</u> the questions.

<u>How many parts</u> are these shapes divided into?   3 parts.

<u>Are</u> these parts equal?   Yes

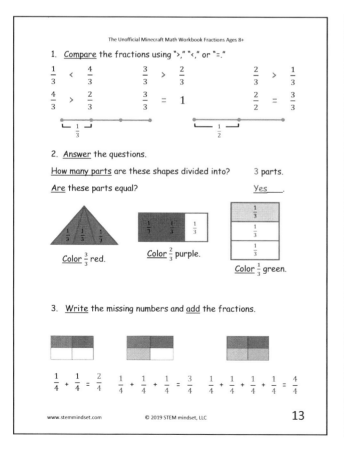

$\frac{1}{3}$ $\frac{1}{3}$ $\frac{1}{3}$

<u>Color</u> $\frac{3}{3}$ red.

$\frac{1}{3}$ $\frac{1}{3}$ $\frac{1}{3}$

<u>Color</u> $\frac{2}{3}$ purple.

$\frac{1}{3}$
$\frac{1}{3}$
$\frac{1}{3}$

<u>Color</u> $\frac{1}{3}$ green.

3. <u>Write</u> the missing numbers and <u>add</u> the fractions.

$\frac{1}{4} + \frac{1}{4} = \frac{2}{4}$ $\frac{1}{4} + \frac{1}{4} + \frac{1}{4} = \frac{3}{4}$ $\frac{1}{4} + \frac{1}{4} + \frac{1}{4} + \frac{1}{4} = \frac{4}{4}$

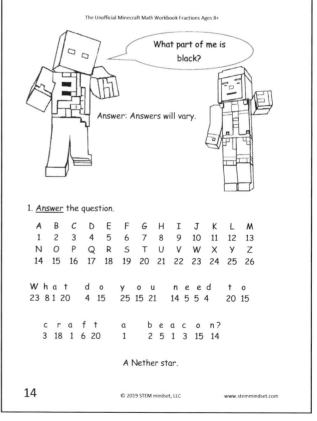

What part of me is black?

Answer: Answers will vary.

1. <u>Answer</u> the question.

A	B	C	D	E	F	G	H	I	J	K	L	M
1	2	3	4	5	6	7	8	9	10	11	12	13
N	O	P	Q	R	S	T	U	V	W	X	Y	Z
14	15	16	17	18	19	20	21	22	23	24	25	26

W h a t   d o   y o u   n e e d   t o
23 8 1 20   4 15   25 15 21   14 5 5 4   20 15

c r a f t   a   b e a c o n?
3 18 1 6 20   1   2 5 1 3 15 14

A Nether star.

1. <u>Add or subtract</u> the fractions. <u>Draw</u> the arrows to match the equation to the picture and to the answer.

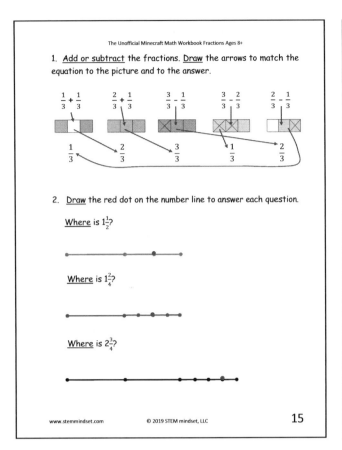

2. <u>Draw</u> the red dot on the number line to answer each question.

<u>Where</u> is $1\frac{1}{2}$?

<u>Where</u> is $1\frac{2}{4}$?

<u>Where</u> is $2\frac{3}{4}$?

---

1. <u>Compare</u> the fractions using ">," "<," or "=."

$$\frac{1}{3} \; < \; \frac{2}{3} \qquad \frac{3}{3} \; > \; \frac{2}{3} \qquad \frac{3}{3} \; = \; 1$$

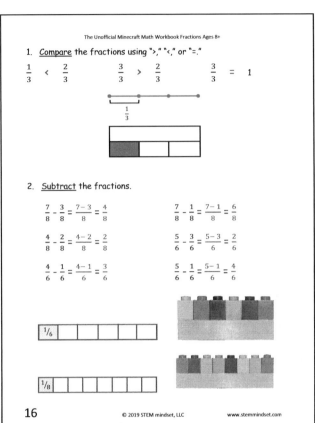

2. <u>Subtract</u> the fractions.

$$\frac{7}{8} - \frac{3}{8} = \frac{7-3}{8} = \frac{4}{8} \qquad\qquad \frac{7}{8} - \frac{1}{8} = \frac{7-1}{8} = \frac{6}{8}$$

$$\frac{4}{8} - \frac{2}{8} = \frac{4-2}{8} = \frac{2}{8} \qquad\qquad \frac{5}{6} - \frac{3}{6} = \frac{5-3}{6} = \frac{2}{6}$$

$$\frac{4}{6} - \frac{1}{6} = \frac{4-1}{6} = \frac{3}{6} \qquad\qquad \frac{5}{6} - \frac{1}{6} = \frac{5-1}{6} = \frac{4}{6}$$

---

1. <u>Answer</u> the questions. Answers may vary.

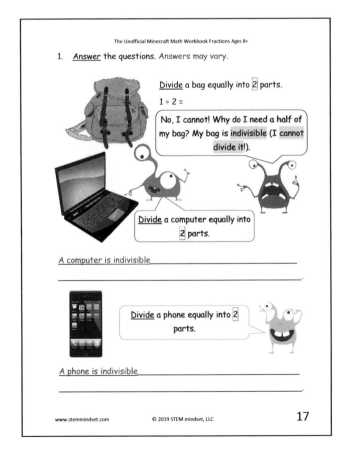

<u>Divide</u> a bag equally into 2 parts.

1 ÷ 2 =

No, I cannot! Why do I need a half of my bag? My bag is indivisible (I cannot divide it!).

<u>Divide</u> a computer equally into 2 parts.

A computer is indivisible _____

_____

<u>Divide</u> a phone equally into 2 parts.

A phone is indivisible _____

_____

What are you thinking of?

Which is longer, one fifth or two fourths?

Answer: two fourths.

80

1. <u>Color</u> $\frac{1}{2}$ *blue,* $\frac{2}{3}$ *yellow,* $\frac{2}{4}$ *red,* $\frac{5}{8}$ *green.*

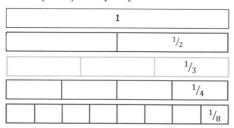

1			
		$^{1}/_{2}$	
		$^{1}/_{3}$	
			$^{1}/_{4}$
			$^{1}/_{8}$

2. <u>Compare</u> the fractions using ">," "<," or "=." It may be helpful to look at the fractions strips above.

$\frac{1}{3}$	>	$\frac{1}{8}$	$\frac{1}{3}$	<	$\frac{3}{8}$
$\frac{1}{2}$	=	$\frac{2}{4}$	$\frac{1}{3}$	>	$\frac{1}{4}$
$\frac{1}{3}$	<	$\frac{1}{2}$	$\frac{3}{4}$	=	$\frac{6}{8}$
$\frac{3}{3}$	>	$\frac{1}{2}$	$\frac{2}{2}$	>	$\frac{7}{8}$
$\frac{2}{4}$	<	$\frac{6}{8}$	$\frac{2}{4}$	>	$\frac{2}{8}$
$\frac{2}{4}$	=	$\frac{4}{8}$	$\frac{2}{3}$	<	$\frac{3}{4}$

---

1. <u>Write</u> the missing numbers and <u>subtract</u> the fractions. The first one is done for you.

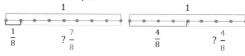

$\frac{1}{8}$? $\frac{7}{8}$ $\frac{4}{8}$? $\frac{4}{8}$

$1 - \frac{1}{8} = \frac{8}{8} - \frac{1}{8} = \frac{8-1}{8} = \frac{7}{8}$ $1 - \frac{4}{8} = \frac{8}{8} - \frac{4}{8} = \frac{8-4}{8} = \frac{4}{8}$

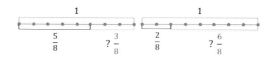

$\frac{5}{8}$? $\frac{3}{8}$ $\frac{2}{8}$? $\frac{6}{8}$

$1 - \frac{5}{8} = \frac{8}{8} - \frac{5}{8} = \frac{8-5}{8} = \frac{3}{8}$ $1 - \frac{2}{8} = \frac{8}{8} - \frac{2}{8} = \frac{8-2}{8} = \frac{6}{8}$

2. <u>Add</u> the fractions and <u>write</u> the value in the rectangles below.

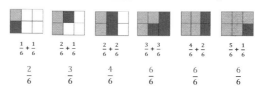

$\frac{1}{6} + \frac{1}{6}$	$\frac{2}{6} + \frac{1}{6}$	$\frac{2}{6} + \frac{2}{6}$	$\frac{3}{6} + \frac{3}{6}$	$\frac{4}{6} + \frac{2}{6}$	$\frac{5}{6} + \frac{1}{6}$
$\frac{2}{6}$	$\frac{3}{6}$	$\frac{4}{6}$	$\frac{6}{6}$	$\frac{6}{6}$	$\frac{6}{6}$

---

1. <u>Write</u> the missing numbers and <u>add</u> the fractions.

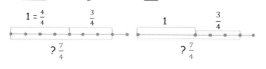

$1 = \frac{4}{4}$ $\frac{3}{4}$ 1 $\frac{3}{4}$

? $\frac{7}{4}$? $\frac{7}{4}$

$1 + \frac{3}{4} = \frac{4}{4} + \frac{3}{4} = \frac{4+3}{4} = \frac{7}{4}$ $1 + \frac{3}{4} = \frac{4}{4} + \frac{3}{4} = \frac{4+3}{4} = \frac{7}{4}$

2. My sister has $\boxed{3}$ hats and she wants to share them equally with her friend. <u>Draw</u> $\boxed{3}$ hats and <u>answer</u> <u>how</u> they can divide the hats equally. *Answers will vary.*

Answer: <u>They can't divide 3 hats equally. Nobody needs a half of</u> <u>a hat.</u>

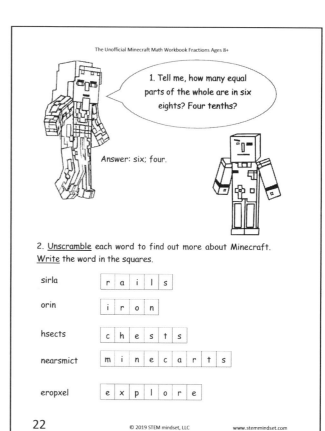

1. Tell me, how many equal parts of the whole are in six eighths? Four tenths?

Answer: six; four.

2. <u>Unscramble</u> each word to find out more about Minecraft. <u>Write</u> the word in the squares.

scrambled	word
sirla	r a i l s
orin	i r o n
hsects	c h e s t s
nearsmict	m i n e c a r t s
eropxel	e x p l o r e

1. Subtract the two fractions. Write the value in the box below.

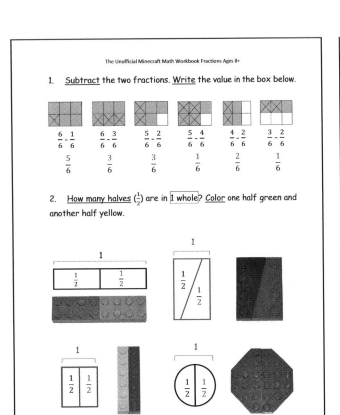

$\frac{6}{6} - \frac{1}{6}$ $\frac{6}{6} - \frac{3}{6}$ $\frac{5}{6} - \frac{2}{6}$ $\frac{5}{6} - \frac{4}{6}$ $\frac{4}{6} - \frac{2}{6}$ $\frac{3}{6} - \frac{2}{6}$

$\frac{5}{6}$ $\frac{3}{6}$ $\frac{3}{6}$ $\frac{1}{6}$ $\frac{2}{6}$ $\frac{1}{6}$

2. How many halves ($\frac{1}{2}$) are in 1 whole? Color one half green and another half yellow.

---

1. Subtract the fractions. Color the minuend yellow and cross out the subtrahend for each problem.

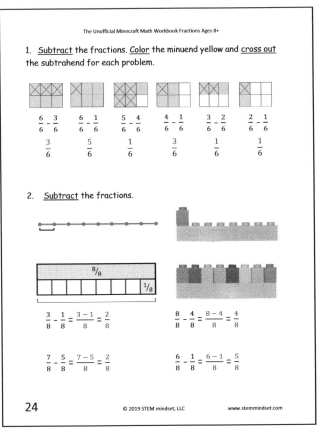

$\frac{6}{6} - \frac{3}{6}$ $\frac{6}{6} - \frac{1}{6}$ $\frac{5}{6} - \frac{4}{6}$ $\frac{4}{6} - \frac{1}{6}$ $\frac{3}{6} - \frac{2}{6}$ $\frac{2}{6} - \frac{1}{6}$

$\frac{3}{6}$ $\frac{5}{6}$ $\frac{1}{6}$ $\frac{3}{6}$ $\frac{1}{6}$ $\frac{1}{6}$

2. Subtract the fractions.

$\frac{3}{8} - \frac{1}{8} = \frac{3-1}{8} = \frac{2}{8}$ $\frac{8}{8} - \frac{4}{8} = \frac{8-4}{8} = \frac{4}{8}$

$\frac{7}{8} - \frac{5}{8} = \frac{7-5}{8} = \frac{2}{8}$ $\frac{6}{8} - \frac{1}{8} = \frac{6-1}{8} = \frac{5}{8}$

---

1. Answer the questions and fill in the missing numbers.

A 5-inch long rectangle is divided into 5 equal parts.

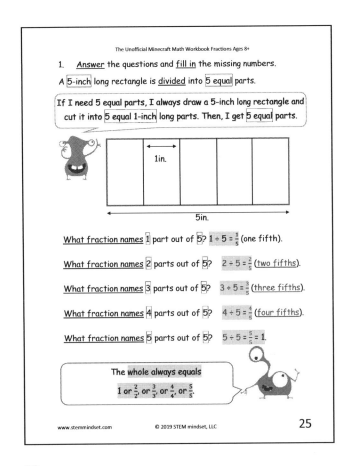

If I need 5 equal parts, I always draw a 5-inch long rectangle and cut it into 5 equal 1-inch long parts. Then, I get 5 equal parts.

1in.

5in.

What fraction names 1 part out of 5? $1 \div 5 = \frac{1}{5}$ (one fifth).

What fraction names 2 parts out of 5? $2 \div 5 = \frac{2}{5}$ (two fifths).

What fraction names 3 parts out of 5? $3 \div 5 = \frac{3}{5}$ (three fifths).

What fraction names 4 parts out of 5? $4 \div 5 = \frac{4}{5}$ (four fifths).

What fraction names 5 parts out of 5? $5 \div 5 = \frac{5}{5} = 1$.

The whole always equals 1 or $\frac{2}{2}$, or $\frac{3}{3}$, or $\frac{4}{4}$, or $\frac{5}{5}$.

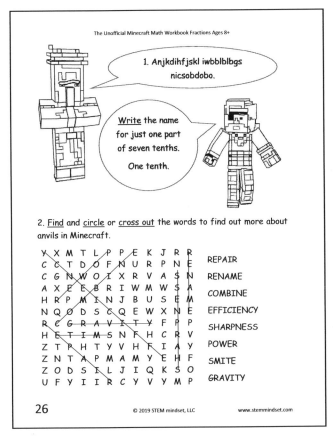

1. Anjkdihfjskl iwbblblbgs nicsobdobo.

Write the name for just one part of seven tenths.

One tenth.

2. Find and circle or cross out the words to find out more about anvils in Minecraft.

```
Y X M T L P P E K J R R
C C T D O F N U R P N E
C G N W O I X R V A S N
A X E E B R I W M W S A
H R P M I N J B U S E M
N Q Ø D S C Q W X N E
R E G R A V I T Y F P P
H E T I M S N F H C R V
Z T P H T Y V H F I A Y
Z N T A P M A M Y E H F
Z O D S I L J I Q K S O
U F Y I I R C Y V Y M P
```

REPAIR

RENAME

COMBINE

EFFICIENCY

SHARPNESS

POWER

SMITE

GRAVITY

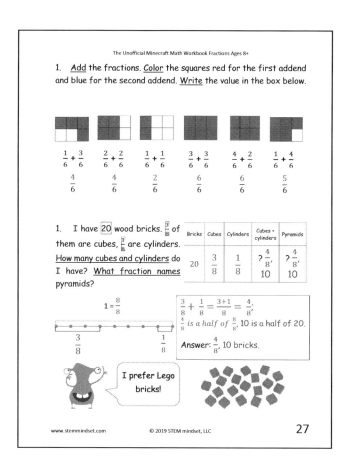

1. <u>Add</u> the fractions. <u>Color</u> the squares red for the first addend and blue for the second addend. <u>Write</u> the value in the box below.

$\frac{1}{6} + \frac{3}{6}$ = $\frac{4}{6}$

$\frac{2}{6} + \frac{2}{6}$ = $\frac{4}{6}$

$\frac{1}{6} + \frac{1}{6}$ = $\frac{2}{6}$

$\frac{3}{6} + \frac{3}{6}$ = $\frac{6}{6}$

$\frac{4}{6} + \frac{2}{6}$ = $\frac{6}{6}$

$\frac{1}{6} + \frac{4}{6}$ = $\frac{5}{6}$

1. I have 20 wood bricks. $\frac{3}{8}$ of them are cubes, $\frac{1}{8}$ are cylinders. <u>How many cubes and cylinders</u> do I have? <u>What fraction names</u> pyramids?

Bricks	Cubes	Cylinders	Cubes + cylinders	Pyramids
20	$\frac{3}{8}$	$\frac{1}{8}$	? $\frac{4}{8}$, 10	? $\frac{4}{8}$, 10

$1 = \frac{8}{8}$

$\frac{3}{8}$ $\frac{1}{8}$

$\frac{3}{8} + \frac{1}{8} = \frac{3+1}{8} = \frac{4}{8}$; $\frac{4}{8}$ is a half of $\frac{8}{8}$, 10 is a half of 20.

Answer: $\frac{4}{8}$, 10 bricks.

I prefer Lego bricks!

Let's look at these triangles. Aha…

We have even numbers – 4, 6, 8, and 10 triangles. Even numbers are numbers that can be divided (split) equally into 2 parts or halves.

1 whole = $\frac{4}{4} = \frac{4\ equal\ parts}{4\ in\ the\ whole}$

2 + 2: $\frac{2}{4} = \frac{1}{2}$ $\frac{2}{4} = \frac{1}{2}$

Aha… I have 4 triangles - 2 parts out of 4 is a half of 4.

And I have 6 triangles - 3 triangles out of 6 is a half of 6 triangles.

1 whole = $\frac{6}{6} = \frac{6\ parts}{6\ whole}$

3 + 3: $\frac{3}{6} = \frac{1}{2}$ $\frac{3}{6} = \frac{1}{2}$

So, I have 8 triangles - 4 triangles out of 8 is a half of 8 triangles.

1 whole = $\frac{8}{8} = \frac{8\ parts}{8\ whole}$

4 + 4: $\frac{4}{8} = \frac{1}{2}$ $\frac{4}{8} = \frac{1}{2}$

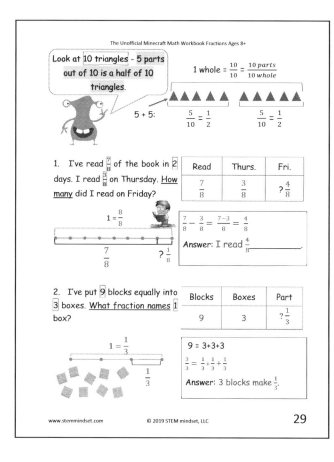

Look at 10 triangles - 5 parts out of 10 is a half of 10 triangles.

1 whole = $\frac{10}{10} = \frac{10\ parts}{10\ whole}$

5 + 5: $\frac{5}{10} = \frac{1}{2}$ $\frac{5}{10} = \frac{1}{2}$

1. I've read $\frac{7}{8}$ of the book in 2 days. I read $\frac{3}{8}$ on Thursday. <u>How many</u> did I read on Friday?

Read	Thurs.	Fri.
$\frac{7}{8}$	$\frac{3}{8}$	? $\frac{4}{8}$

$1 = \frac{8}{8}$

$\frac{7}{8}$? $\frac{1}{8}$

$\frac{7}{8} - \frac{3}{8} = \frac{7-3}{8} = \frac{4}{8}$

Answer: I read $\frac{4}{8}$

2. I've put 9 blocks equally into 3 boxes. <u>What fraction names</u> 1 box?

Blocks	Boxes	Part
9	3	? $\frac{1}{3}$

$1 = \frac{1}{3}$

$\frac{1}{3}$

$9 = 3+3+3$

$\frac{3}{3} = \frac{1}{3} + \frac{1}{3} + \frac{1}{3}$

Answer: 3 blocks make $\frac{1}{3}$.

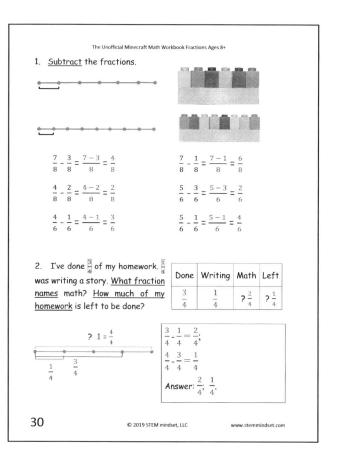

1. <u>Subtract</u> the fractions.

$\frac{7}{8} - \frac{3}{8} = \frac{7-3}{8} = \frac{4}{8}$ $\frac{7}{8} - \frac{1}{8} = \frac{7-1}{8} = \frac{6}{8}$

$\frac{4}{8} - \frac{2}{8} = \frac{4-2}{8} = \frac{2}{8}$ $\frac{5}{6} - \frac{3}{6} = \frac{5-3}{6} = \frac{2}{6}$

$\frac{4}{6} - \frac{1}{6} = \frac{4-1}{6} = \frac{3}{6}$ $\frac{5}{6} - \frac{1}{6} = \frac{5-1}{6} = \frac{4}{6}$

2. I've done $\frac{3}{4}$ of my homework. $\frac{1}{4}$ was writing a story. <u>What fraction names</u> math? <u>How much of my homework is left</u> to be done?

Done	Writing	Math	Left
$\frac{3}{4}$	$\frac{1}{4}$	? $\frac{2}{4}$	? $\frac{1}{4}$

? $1 = \frac{4}{4}$

$\frac{1}{4}$ $\frac{3}{4}$

$\frac{3}{4} - \frac{1}{4} = \frac{2}{4}$;

$\frac{4}{4} - \frac{3}{4} = \frac{1}{4}$

Answer: $\frac{2}{4}$, $\frac{1}{4}$.

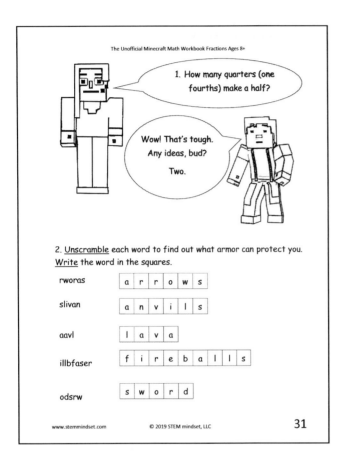

1. How many quarters (one fourths) make a half?

Wow! That's tough. Any ideas, bud? Two.

2. Unscramble each word to find out what armor can protect you. Write the word in the squares.

rworas	a	r	r	o	w	s
slivan	a	n	v	i	l	s
aavl	l	a	v	a
illbfaser	f	i	r	e	b	a	l	l	s
odsrw	s	w	o	r	d

---

1. Write the missing numbers to make the fractions complete.

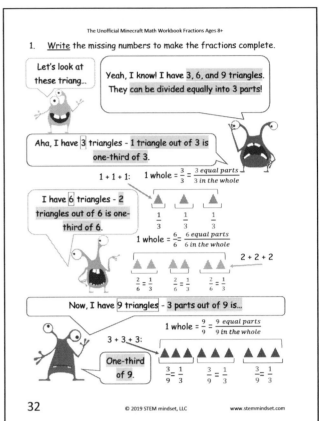

Let's look at these triang...

Yeah, I know! I have 3, 6, and 9 triangles. They can be divided equally into 3 parts!

Aha, I have 3 triangles - 1 triangle out of 3 is one-third of 3.

$1 + 1 + 1$: $\quad$ 1 whole $= \frac{3}{3} = \frac{3 \ equal \ parts}{3 \ in \ the \ whole}$

$\frac{1}{3} \quad \frac{1}{3} \quad \frac{1}{3}$

I have 6 triangles - 2 triangles out of 6 is one-third of 6.

1 whole $= \frac{6}{6} = \frac{6 \ equal \ parts}{6 \ in \ the \ whole}$

$2 + 2 + 2$

$\frac{2}{6} = \frac{1}{3} \quad \frac{2}{6} = \frac{1}{3} \quad \frac{2}{6} = \frac{1}{3}$

Now, I have 9 triangles - 3 parts out of 9 is...

1 whole $= \frac{9}{9} = \frac{9 \ equal \ parts}{9 \ in \ the \ whole}$

$3 + 3 + 3$:

One-third of 9.

$\frac{3}{9} = \frac{1}{3} \quad \frac{3}{9} = \frac{1}{3} \quad \frac{3}{9} = \frac{1}{3}$

---

1. Answer the questions by filling in the missing numbers.

A 3-inch long square is divided into 3 equal parts each 1-inch long.

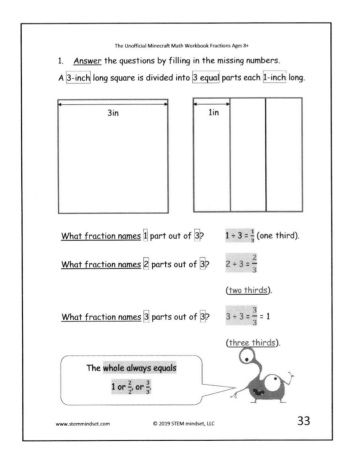

3in

1in

What fraction names 1 part out of 3? $\quad 1 \div 3 = \frac{1}{3}$ (one third).

What fraction names 2 parts out of 3? $\quad 2 \div 3 = \frac{2}{3}$

(two thirds).

What fraction names 3 parts out of 3? $\quad 3 \div 3 = \frac{3}{3} = 1$

(three thirds).

The whole always equals 1 or $\frac{2}{2}$ or $\frac{3}{3}$.

---

1. I've had 5 apples equally for lunch in 5 days. What part of the apples have I eaten in 3 days?

Apples	Days	Part
5	3	? $\frac{3}{5}$

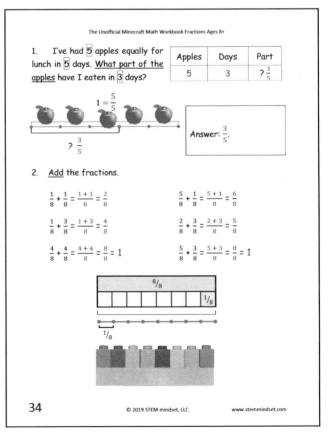

$1 = \frac{5}{5}$

? $\frac{3}{5}$

Answer: $\frac{3}{5}$.

2. Add the fractions.

$\frac{1}{8} + \frac{1}{8} = \frac{1+1}{8} = \frac{2}{8}$ $\qquad$ $\frac{5}{8} + \frac{1}{8} = \frac{5+1}{8} = \frac{6}{8}$

$\frac{1}{8} + \frac{3}{8} = \frac{1+3}{8} = \frac{4}{8}$ $\qquad$ $\frac{2}{8} + \frac{3}{8} = \frac{2+3}{8} = \frac{5}{8}$

$\frac{4}{8} + \frac{4}{8} = \frac{4+4}{8} = \frac{8}{8} = 1$ $\qquad$ $\frac{5}{8} + \frac{3}{8} = \frac{5+3}{8} = \frac{8}{8} = 1$

$^8/_8$

$^1/_8$

$^1/_8$

1. I'm less than 4 wholes. I'm made of 5 halves. Who am I?

Easy! I'm ... Hm.... I desperately need your help, BFF! Two and a half.

2. <u>Answer</u> the question.

A	B	C	D	E	F	G	H	I	J	K	L	M
1	2	3	4	5	6	7	8	9	10	11	12	13
N	O	P	Q	R	S	T	U	V	W	X	Y	Z
14	15	16	17	18	19	20	21	22	23	24	25	26

W h a t   d o   y o u   n e e d   t o
23 8 1 20   4 15   25 15 21   14 5 5 4   20 15

c r a f t   a   s e t   o f   a r m o r?
3 18 1 20   1   19 5 20   15 6   1 18 13 15 18

Twenty-four pieces of the same material.

---

1. <u>What fraction</u> is more? Use ">," "<," or "=."

$\frac{1}{2}$ > $\frac{1}{4}$ $\frac{1}{8}$ < $\frac{3}{4}$

$\frac{1}{8}$ < $\frac{1}{2}$ $\frac{1}{2}$ = $\frac{2}{4}$

$\frac{3}{4}$ < $\frac{7}{8}$ $\frac{5}{8}$ > $\frac{1}{2}$

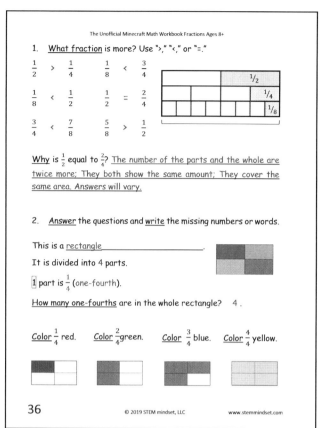

<u>Why</u> is $\frac{1}{2}$ equal to $\frac{2}{4}$? <u>The number of the parts and the whole are twice more; They both show the same amount; They cover the same area. Answers will vary.</u>

2. <u>Answer</u> the questions and <u>write</u> the missing numbers or words.

This is a <u>rectangle</u>.
It is divided into 4 parts.
1 part is $\frac{1}{4}$ (one-fourth).
<u>How many one-fourths</u> are in the whole rectangle?   4 .

<u>Color</u> $\frac{1}{4}$ red.   <u>Color</u> $\frac{2}{4}$ green.   <u>Color</u> $\frac{3}{4}$ blue.   <u>Color</u> $\frac{4}{4}$ yellow.

---

1. I put 15 cupcakes equally on 3 plates. <u>What part of the cupcakes</u> was on 1 plate? <u>Fill in</u> the missing numbers.

Cupcakes	Plates	Part
15	3	? $\frac{1}{3}$

$1 = \frac{3}{3}$

$\frac{1}{3}$

To solve the problem, I use a diagram. See, I have 3 plates, I make 3 line segments, I can divide 15 cupcakes equally into 3 parts: 15 = 5 + 5 + 5. Since 1 part out of 3 plates equals $\frac{1}{3}$, 5 cupcakes on 1 plates equal $\frac{1}{3}$, too.

$\frac{3}{3} = 15$

$\frac{1}{3} = 5$

Answer: $\frac{1}{3}$

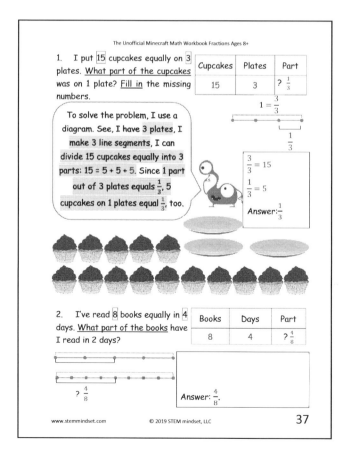

2. I've read 8 books equally in 4 days. <u>What part of the books</u> have I read in 2 days?

Books	Days	Part
8	4	? $\frac{4}{8}$

? $\frac{4}{8}$

Answer: $\frac{4}{8}$.

---

1. <u>Fill in</u> the missing numbers to make the fractions complete.

Let's look at these triangles and think.

I can divide numbers 4 and 8 into 4 equal parts and 5 and 10 into 5 equal parts.

1 + 1 + 1 + 1:
1 whole = $\frac{4}{4} = \frac{4\ parts}{4\ whole}$

So, 1 part out of 4 is a one fourth of 4.

$\frac{1}{4}$ $\frac{1}{4}$ $\frac{1}{4}$ $\frac{1}{4}$

2 + 2 + 2 + 2:
1 whole = $\frac{8}{8} = \frac{8\ parts}{8\ whole}$

I know that 2 parts out of 8 is one fourth of 8.

$\frac{2}{8} = \frac{1}{4}$ $\frac{2}{8} = \frac{1}{4}$ $\frac{2}{8} = \frac{1}{4}$ $\frac{2}{8} = \frac{1}{4}$

I know that 2 triangles out of 10 is one fifth of 10.

2 + 2 + 2 + 2 + 2:
1 whole = $\frac{10}{10} = \frac{10\ parts}{10\ whole}$

$\frac{2}{10} = \frac{1}{5}$ $\frac{2}{10} = \frac{1}{5}$ $\frac{2}{10} = \frac{1}{5}$ $\frac{2}{10} = \frac{1}{5}$ $\frac{2}{10} = \frac{1}{5}$

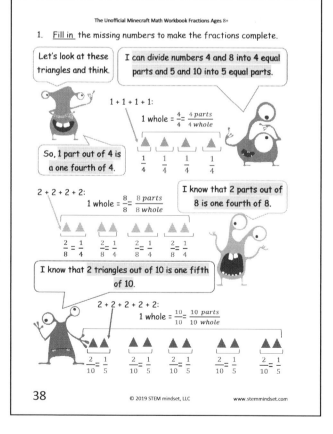

1. <u>Compare</u> the fractions using the fractions strips (">," "<," or "=.")

$$\frac{1}{2} > \frac{1}{5}$$

$$\frac{1}{8} < \frac{2}{5}$$

$$\frac{2}{4} < \frac{3}{5}$$

$$\frac{1}{2} > \frac{2}{5}$$

$$\frac{2}{4} = \frac{4}{8}$$

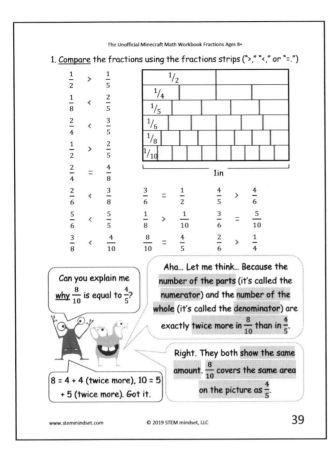

$$\frac{2}{6} < \frac{3}{8} \qquad \frac{3}{6} = \frac{1}{2} \qquad \frac{4}{5} > \frac{4}{6}$$

$$\frac{5}{6} < \frac{5}{5} \qquad \frac{1}{8} > \frac{1}{10} \qquad \frac{3}{6} = \frac{5}{10}$$

$$\frac{3}{8} < \frac{4}{10} \qquad \frac{8}{10} = \frac{4}{5} \qquad \frac{2}{6} > \frac{1}{4}$$

Can you explain me why $\frac{8}{10}$ is equal to $\frac{4}{5}$?

Aha... Let me think... Because the number of the parts (it's called the numerator) and the number of the whole (it's called the denominator) are exactly twice more in $\frac{8}{10}$ than in $\frac{4}{5}$.

Right. They both show the same amount. $\frac{8}{10}$ covers the same area on the picture as $\frac{4}{5}$.

8 = 4 + 4 (twice more), 10 = 5 + 5 (twice more). Got it.

1. My denominator is even and less than 15. My numerator is odd and less than 10. I'm equivalent to one fourth. Who am I?

I know the denominator may be 2, 4, 6, 8, 10, 12, and 14. The numerator may be 1, 3, 5, 7, and 9. I'm almost there, can you help me? Three twelfths.

---

1. <u>Answer</u> the questions and <u>fill in</u> the missing numbers.

A rectangle is divided into 9 equal parts. Hints: 1) <u>divide</u> a rectangle into 3 equal parts (black lines); 2) <u>divide</u> each remaining part into 3 equal parts (blue-dotted lines). You will get 9 equal parts.

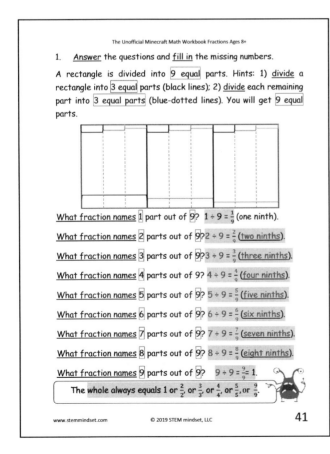

<u>What fraction names</u> 1 part out of 9? $1 \div 9 = \frac{1}{9}$ (one ninth).

<u>What fraction names</u> 2 parts out of 9? $2 \div 9 = \frac{2}{9}$ (two ninths).

<u>What fraction names</u> 3 parts out of 9? $3 \div 9 = \frac{3}{9}$ (three ninths).

<u>What fraction names</u> 4 parts out of 9? $4 \div 9 = \frac{4}{9}$ (four ninths).

<u>What fraction names</u> 5 parts out of 9? $5 \div 9 = \frac{5}{9}$ (five ninths).

<u>What fraction names</u> 6 parts out of 9? $6 \div 9 = \frac{6}{9}$ (six ninths).

<u>What fraction names</u> 7 parts out of 9? $7 \div 9 = \frac{7}{9}$ (seven ninths).

<u>What fraction names</u> 8 parts out of 9? $8 \div 9 = \frac{8}{9}$ (eight ninths).

<u>What fraction names</u> 9 parts out of 9? $9 \div 9 = \frac{9}{9} = 1$.

The whole always equals 1 or $\frac{2}{2}$, or $\frac{3}{3}$, or $\frac{4}{4}$, or $\frac{5}{5}$, or $\frac{9}{9}$.

---

1. <u>Write</u> the missing numbers and <u>subtract</u> the fractions. The first one is done for you.

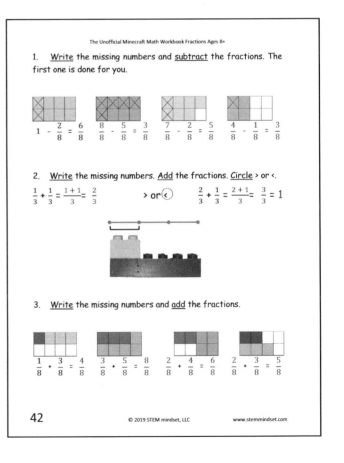

$$1 - \frac{2}{8} = \frac{6}{8} \qquad \frac{8}{8} - \frac{5}{8} = \frac{3}{8} \qquad \frac{7}{8} - \frac{2}{8} = \frac{5}{8} \qquad \frac{4}{8} - \frac{1}{8} = \frac{3}{8}$$

2. <u>Write</u> the missing numbers. <u>Add</u> the fractions. <u>Circle</u> > or <.

$$\frac{1}{3} + \frac{1}{3} = \frac{1+1}{3} = \frac{2}{3} \qquad > \text{ or } < \qquad \frac{2}{3} + \frac{1}{3} = \frac{2+1}{3} = \frac{3}{3} = 1$$

3. <u>Write</u> the missing numbers and <u>add</u> the fractions.

$$\frac{1}{8} + \frac{3}{8} = \frac{4}{8} \qquad \frac{3}{8} + \frac{5}{8} = \frac{8}{8} \qquad \frac{2}{8} + \frac{4}{8} = \frac{6}{8} \qquad \frac{2}{8} + \frac{3}{8} = \frac{5}{8}$$

1. Subtract and compare (">" or "<") the fractions.

$$1 - \frac{5}{6} = \frac{6}{6} - \frac{5}{6} = \frac{6-5}{6} = \frac{1}{6} \quad < \quad 1 - \frac{2}{3} = \frac{3}{3} - \frac{2}{3} = \frac{3-2}{3} = \frac{1}{3}$$

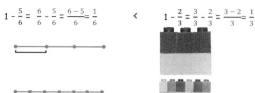

2. Write the missing numbers and subtract the fractions. The first one is done for you.

1	1
$\frac{1}{2}$? $\frac{1}{2}$	$\frac{1}{4}$? $\frac{3}{4}$

$1 - \frac{1}{2} = \frac{2}{2} - \frac{1}{2} = \frac{2-1}{2} = \frac{1}{2}$ $\qquad$ $1 - \frac{1}{4} = \frac{4}{4} - \frac{1}{4} = \frac{4-1}{4} = \frac{3}{4}$

1	1
$\frac{2}{4}$? $\frac{2}{4}$	$\frac{3}{4}$? $\frac{1}{4}$

$1 - \frac{2}{4} = \frac{4}{4} - \frac{2}{4} = \frac{4-2}{4} = \frac{2}{4}$ $\qquad$ $1 - \frac{3}{4} = \frac{4}{4} - \frac{3}{4} = \frac{4-3}{4} = \frac{1}{4}$

---

1. Answer the questions and write the missing numbers.

A rectangle is divided into 8 equal parts. Hints: 1) divide a rectangle into 2 equal parts (blue-dotted line); 2) divide each half into 2 equal parts (red-dotted lines); 3) divide each quarter into 2 equal parts (black and green-dotted lines). You will get 8 equal parts altogether.

What fraction names 1 part out of 8? $1 \div 8 = \frac{1}{8}$ (one eighth).

How many are 2 parts out of 8? $2 \div 8 = \frac{2}{8}$ (two eighths).

What fraction names 3 parts out of 8? $3 \div 8 = \frac{3}{8}$ (three eighths).

What fraction names 4 parts out of 8? $4 \div 8 = \frac{4}{8}$ (four eighths).

4 parts equal a half of the rectangle. So, $\frac{4}{8} = \frac{1}{2}$.

What fraction names 5 parts out of 8? $5 \div 8 = \frac{5}{8}$ (five eighths).

What fraction names 6 parts out of 8? $6 \div 8 = \frac{6}{8}$ (six eighths).

What fraction names 7 parts out of 8? $7 \div 8 = \frac{7}{8}$ (seven eighths).

What fraction names 8 parts out of 8? $8 \div 8 = \frac{8}{8} = 1$.

The whole always equals 1 or $\frac{2}{2}$, or $\frac{3}{3}$, or $\frac{4}{4}$, or $\frac{5}{5}$, $\frac{8}{8}$.

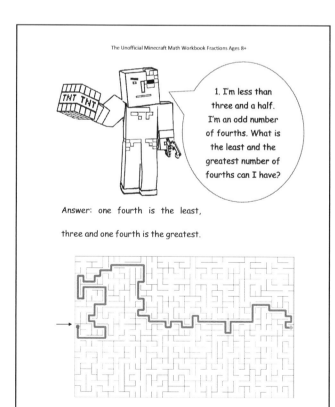

1. I'm less than three and a half. I'm an odd number of fourths. What is the least and the greatest number of fourths can I have?

Answer: one fourth is the least,

three and one fourth is the greatest.

---

1. Add and compare the fractions. Circle ">," "<," or "=."

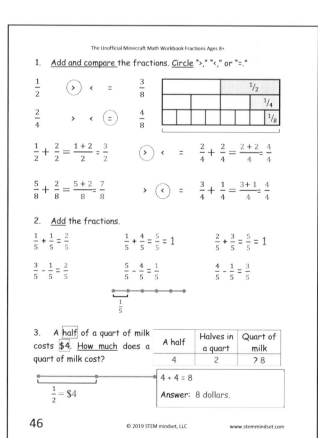

$\frac{1}{2}$ (>) < = $\frac{3}{8}$

$\frac{2}{4}$ > < (=) $\frac{4}{8}$

$\frac{1}{2} + \frac{2}{2} = \frac{1+2}{2} = \frac{3}{2}$ (>) < = $\frac{2}{4} + \frac{2}{4} = \frac{2+2}{4} = \frac{4}{4}$

$\frac{5}{8} + \frac{2}{8} = \frac{5+2}{8} = \frac{7}{8}$ > (<) = $\frac{3}{4} + \frac{1}{4} = \frac{3+1}{4} = \frac{4}{4}$

2. Add the fractions.

$\frac{1}{5} + \frac{1}{5} = \frac{2}{5}$ $\qquad$ $\frac{1}{5} + \frac{4}{5} = \frac{5}{5} = 1$ $\qquad$ $\frac{2}{5} + \frac{3}{5} = \frac{5}{5} = 1$

$\frac{3}{5} - \frac{1}{5} = \frac{2}{5}$ $\qquad$ $\frac{5}{5} - \frac{4}{5} = \frac{1}{5}$ $\qquad$ $\frac{4}{5} - \frac{1}{5} = \frac{3}{5}$

$\frac{1}{5}$

3. A half of a quart of milk costs $4. How much does a quart of milk cost?

A half	Halves in a quart	Quart of milk
4	2	? 8

$\frac{1}{2} = \$4$

$4 + 4 = 8$

Answer: 8 dollars.

1. <u>Answer</u> the questions and <u>fill in</u> the missing numbers.

A rectangle is <u>divided</u> into 4 equal parts.

Hints: 1) <u>divide</u> a sheet of paper into 2 equal parts (blue-dotted line); 2) <u>divide</u> each <u>half</u> into 2 equal parts (red-dotted lines). You will get 4 equal parts in all.

What fractions names 1 part out of 4 equal parts?

$1 \div 4 = \frac{1}{4}$ (one fourth).

<u>What fractions names</u> 2 parts out of 4?

$2 \div 4 = \frac{2}{4}$ (two fourths).

2 parts equal a half of the rectangle. So,

$$\frac{2}{4} = \frac{1}{2}.$$

<u>What fractions names</u> 3 parts out of 4? $3 \div 4 = \frac{3}{4}$

or 2 parts + 1 part: $\frac{2}{4} + \frac{1}{4} = \frac{3}{4}$ (three fourths).

<u>What fractions names</u> 4 parts out of 4? $4 \div 4 = \frac{4}{4} = 1$.

The whole always equals 1 or $\frac{2}{2}$, or $\frac{3}{3}$, or $\frac{4}{4}$.

---

1. <u>Subtract and compare</u> the fractions (<u>use</u> ">" or "<").

$\frac{5}{5} - \frac{1}{5} = \frac{4}{5}$ > $\frac{4}{5} - \frac{1}{5} = \frac{3}{5}$ > $\frac{3}{5} - \frac{2}{5} = \frac{1}{5}$

$\frac{3}{5} + \frac{1}{5} - \frac{2}{5} = \frac{2}{5}$ < $\frac{4}{5} - \frac{2}{5} + \frac{3}{5} = \frac{5}{5}$ > $\frac{2}{5} + \frac{3}{5} - \frac{4}{5} = \frac{1}{5}$

$\frac{1}{5}$

2. I have 10 halves of a cup. <u>How many cups</u> do I have in the pot?

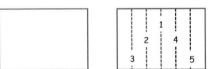

I can add halves, or I can add cups.

1 cup = 2 halves of a cup. So, 1 cup = $\frac{1}{2} + \frac{1}{2} = \frac{1+1}{2} = \frac{2}{2} = 1$

1 + 1 + 1 + 1 + 1 = 5

Answer: 5 cups_____.

1. In a group of like-fractions of equal blocks how do you tell which fraction is largest?

Answer: the fraction with the greatest numerator.

2. <u>Find</u> and <u>circle</u> or <u>cross out</u> the words to find out more about natural landscapes in Minecraft.

```
L R G T Z S N D R E A R       BIOMES
K L R O E C Q Q F E S E
W D A M L Q Y N L G R H        TEMPERATURE
U L O F Z D R U A Y Z T
L I G T N I E D C I G A        WEATHER
B X J K V I A R P W C E
B H U P J Z A G A A D W        SNOWY
V F V X H X Q R W R K U        COLD
N X J W Q Q D F O M D Q
E R U T A R E P M E T X        DRY
S N O W Y O H E O Z X X        WARM
B Z Y Z O I K L X J M T        RAINFALL
```

---

1. <u>Answer</u> the questions and <u>fill in</u> the missing numbers.

Below is a sheet of paper <u>divided</u> into 6 equal parts. Hints: 1) <u>divide</u> a sheet of paper into 2 equal parts (green-dotted line); 2) <u>divide</u> each half into 3 equal parts (red-dotted lines, then, black-dotted lines). You will get 6 equal parts in all.

<u>What fraction names</u> 1 part out of 6? $1 \div 6 = \frac{1}{6}$ (one sixth).

<u>What fraction names</u> 2 parts out of 6? $2 \div 6 = \frac{2}{6}$ (two sixths).

<u>What fraction names</u> 3 parts out of 6? $3 \div 6 = \frac{3}{6}$ (three sixths).

3 parts equal a half of the rectangle. So, $\frac{3}{6} = \frac{1}{2}$.

<u>What fraction names</u> 4 parts out of 6? $4 \div 6 = \frac{4}{6}$ (four sixths).

<u>What fraction names</u> 5 parts out of 6? $5 \div 6 = \frac{5}{6}$ (five sixths).

The whole always equals 1 or $\frac{2}{2}$, or $\frac{3}{3}$, or $\frac{4}{4}$, or $\frac{6}{6}$. etc.

1. <u>Compare</u> the fractions. <u>Use</u> "=," ">," or "<" and the fractions strips below.

$\frac{3}{4}$ > $\frac{3}{8}$ $\frac{2}{8}$ > $\frac{2}{12}$

$\frac{1}{4}$ = $\frac{2}{8}$ $\frac{2}{4}$ > $\frac{4}{12}$

$\frac{9}{12}$ = $\frac{3}{4}$ $\frac{2}{8}$ = $\frac{3}{12}$

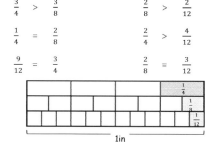

1in

2. <u>Answer</u> the questions.

How many parts are in this bar? 4 parts.

<u>What fraction names</u> 1 block? $\frac{1}{4}$ $\frac{1}{4}$ 8cm

How long is $\frac{1}{4}$ if the bar is 8 cm long?   2 cm.

How long is $\frac{2}{4}$ if the bar is 8 cm long? _____   4 cm.

How long is $\frac{3}{4}$ if the bar is 8 cm long? _____   6 cm.

---

1. <u>Answer</u> the questions.

<u>How many parts</u> are in this bar?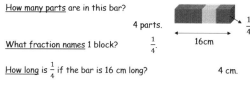

4 parts. $\frac{1}{4}$

<u>What fraction names</u> 1 block? $\frac{1}{4}$.   16cm

<u>How long</u> is $\frac{1}{4}$ if the bar is 16 cm long?   4 cm.

<u>How long</u> is $\frac{2}{4}$ if the bar is 16 cm long? _____   8 cm.

<u>How long</u> is $\frac{3}{4}$ if the bar is 16 cm long? _____   12 cm.

2. <u>How many more</u> do you need to add to make 1 whole? <u>Write</u> the missing fraction.

$\frac{1}{2}+\frac{1}{2}$ $\frac{3}{4}+\frac{1}{4}$ $\frac{5}{6}+\frac{1}{6}$ $\frac{2}{3}+\frac{1}{3}$ $\frac{9}{10}+\frac{1}{10}$ $\frac{4}{5}+\frac{1}{5}$ $\frac{7}{8}+\frac{1}{8}$

3. <u>Write</u> the missing fractions.

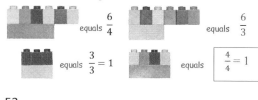

equals $\frac{6}{4}$ equals $\frac{6}{3}$

equals $\frac{3}{3}=1$ equals $\boxed{\frac{4}{4}=1}$

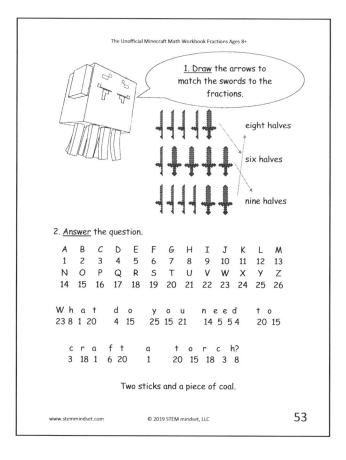

1. <u>Draw</u> the arrows to match the swords to the fractions.

eight halves

six halves

nine halves

2. <u>Answer</u> the question.

A	B	C	D	E	F	G	H	I	J	K	L	M
1	2	3	4	5	6	7	8	9	10	11	12	13
N	O	P	Q	R	S	T	U	V	W	X	Y	Z
14	15	16	17	18	19	20	21	22	23	24	25	26

W h a t   d o   y o u   n e e d   t o
23 8 1 20   4 15   25 15 21   14 5 5 4   20 15

c r a f t   a   t o r c h?
3 18 1 6 20   1   20 15 18 3 8

Two sticks and a piece of coal.

---

1. <u>Answer</u> the questions and <u>fill in</u> the missing numbers.

<u>How many parts</u> are in this bar?   8 parts.

<u>What fraction names</u> 1 block? $\frac{1}{8}$.

<u>How long</u> is $\frac{1}{8}$ if the bar is 80 cm long?   10 cm.

<u>How</u> did you find it?   10 + 10 + 10 + 10 + 10 + 10 + 10 + 10 = 80.

<u>How long</u> is $\frac{3}{8}$?   30 cm.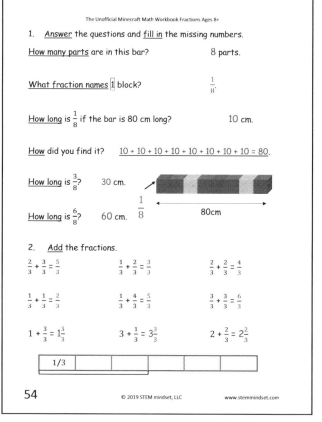

<u>How long</u> is $\frac{6}{8}$?   60 cm. $\frac{1}{8}$ 80cm

2. <u>Add</u> the fractions.

$\frac{2}{3}+\frac{3}{3}=\frac{5}{3}$ $\frac{1}{3}+\frac{2}{3}=\frac{3}{3}$ $\frac{2}{3}+\frac{2}{3}=\frac{4}{3}$

$\frac{1}{3}+\frac{1}{3}=\frac{2}{3}$ $\frac{1}{3}+\frac{4}{3}=\frac{5}{3}$ $\frac{3}{3}+\frac{3}{3}=\frac{6}{3}$

$1+\frac{3}{3}=1\frac{3}{3}$ $3+\frac{1}{3}=3\frac{3}{3}$ $2+\frac{2}{3}=2\frac{2}{3}$

1/3				

1. <u>Write</u> the missing numbers.

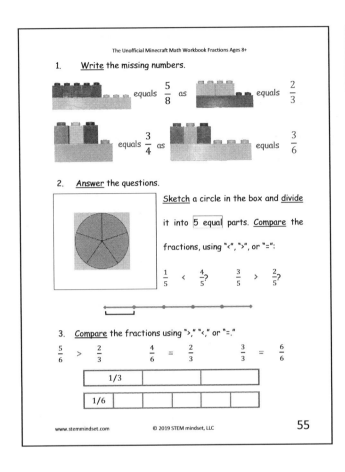

equals $\frac{5}{8}$ as equals $\frac{2}{3}$

equals $\frac{3}{4}$ as equals $\frac{3}{6}$

2. <u>Answer</u> the questions.

<u>Sketch</u> a circle in the box and <u>divide</u> it into $\boxed{5}$ equal parts. <u>Compare</u> the fractions, using "<", ">", or "=":

$\frac{1}{5}$ < $\frac{4}{5}$? $\frac{3}{5}$ > $\frac{2}{5}$?

3. <u>Compare</u> the fractions using ">," "<," or "=."

$\frac{5}{6}$ > $\frac{2}{3}$ $\frac{4}{6}$ = $\frac{2}{3}$ $\frac{3}{3}$ = $\frac{6}{6}$

1/3		

1/6					

---

1. <u>Evaluate</u> each equation.

$1 - \frac{2}{3} = \frac{1}{3}$ $1 - \frac{1}{3} = \frac{2}{3}$ $1 - \frac{3}{3} = \frac{0}{3} = 0$

$\frac{2}{3} - \frac{1}{3} = \frac{1}{3}$ $\frac{4}{3} - \frac{1}{3} = \frac{3}{3} = 1$ $\frac{3}{3} - \frac{2}{3} = \frac{1}{3}$

$\frac{3}{3} + \frac{1}{3} - \frac{2}{3} = \frac{2}{3}$ $\frac{4}{3} - \frac{2}{3} - \frac{1}{3} = \frac{1}{3}$ $\frac{2}{3} + \frac{3}{3} - \frac{4}{3} = \frac{1}{3}$

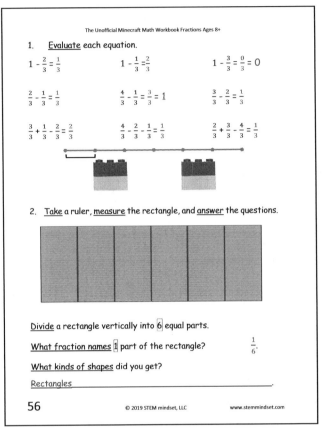

2. <u>Take</u> a ruler, <u>measure</u> the rectangle, and <u>answer</u> the questions.

<u>Divide</u> a rectangle vertically into $\boxed{6}$ equal parts.

<u>What fraction names</u> $\boxed{1}$ part of the rectangle? $\frac{1}{6}$.

<u>What kinds of shapes</u> did you get?

Rectangles   .

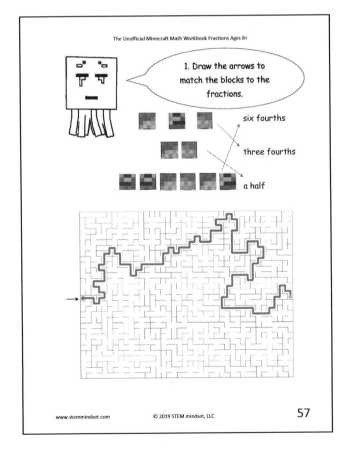

1. Draw the arrows to match the blocks to the fractions.

six fourths

three fourths

a half

---

1. <u>Compare</u> the fractions using ">," "<," or "=."

$\frac{1}{2}$ < $\frac{4}{6}$ $\frac{1}{6}$ < $\frac{1}{3}$ $\frac{2}{3}$ = $\frac{4}{6}$

1/6					
1/3			
1/2		

2. <u>Add or subtract</u> the fractions.

$\frac{1}{6} + \frac{2}{6} = \frac{3}{6}$ $\frac{1}{6} + \frac{3}{6} = \frac{4}{6}$ $\frac{4}{6} + \frac{1}{6} = \frac{5}{6}$

$1 + \frac{5}{6} = 1\frac{5}{6}$ $3 + \frac{1}{6} = 3\frac{1}{6}$ $2 + \frac{2}{6} = 2\frac{2}{6}$

$1 - \frac{2}{6} = \frac{4}{6}$ $1 - \frac{1}{6} = \frac{5}{6}$ $1 - \frac{3}{6} = \frac{3}{6}$

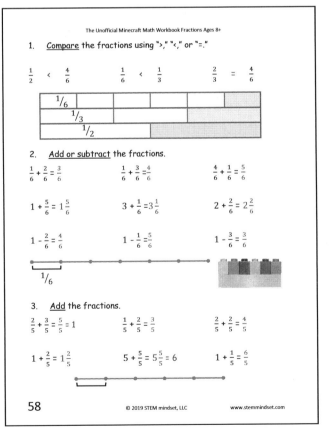

1/6

3. <u>Add</u> the fractions.

$\frac{2}{5} + \frac{3}{5} = \frac{5}{5} = 1$ $\frac{1}{5} + \frac{2}{5} = \frac{3}{5}$ $\frac{2}{5} + \frac{2}{5} = \frac{4}{5}$

$1 + \frac{2}{5} = 1\frac{2}{5}$ $5 + \frac{5}{5} = 5\frac{5}{5} = 6$ $1 + \frac{1}{5} = \frac{6}{5}$

90

1. Add and subtract the fractions.

$\frac{2}{6} - \frac{1}{6} = \frac{1}{6}$ $\frac{4}{6} - \frac{1}{6} = \frac{3}{6}$ $\frac{2}{6} + \frac{3}{6} - \frac{4}{6} = \frac{1}{6}$

$\frac{3}{6} + \frac{1}{6} - \frac{2}{6} = \frac{2}{6}$ $\frac{4}{6} - \frac{2}{6} - \frac{1}{6} = \frac{1}{6}$ $\frac{3}{6} - \frac{2}{6} = \frac{1}{6}$

$^6/_6$

$^3/_6$

2. Compare fractions using the fractions strips below, fill in ">,"

"<," or "=."

$\frac{1}{3}$ = $\frac{2}{6}$ $\frac{4}{6}$ < $\frac{3}{3}$

$\frac{1}{3}$ < $\frac{1}{2}$ $\frac{5}{6}$ > $\frac{3}{6}$

$^1/_6$					
$^1/_3$			
$^1/_2$		

---

1. Write the missing fraction.

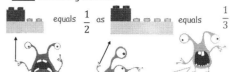

equals $\frac{1}{2}$ as   equals $\frac{1}{3}$

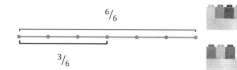

Hmm... Fractions?!

2. Add and subtract the fractions.

$\frac{4}{6} + \frac{2}{6} = \frac{6}{6} = 1$ $\frac{2}{6} + \frac{3}{6} = \frac{5}{6}$ $\frac{1}{6} + \frac{1}{6} = \frac{2}{6}$

$1 + \frac{3}{6} = 1\frac{3}{6}$ $3 + \frac{6}{6} = 3\frac{6}{6} = 4$ $2 + \frac{4}{6} = 2\frac{4}{6}$

$2 - \frac{6}{6} = 1$ $1 - \frac{4}{6} = \frac{2}{6}$ $1 - \frac{5}{6} = \frac{1}{6}$

3. Evaluate each equation.

$\frac{5}{5} - \frac{1}{5} = \frac{4}{5}$ $\frac{4}{5} - \frac{1}{5} = \frac{3}{5}$ $\frac{3}{5} - \frac{2}{5} = \frac{1}{5}$

$\frac{3}{5} + \frac{1}{5} - \frac{2}{5} = \frac{2}{5}$ $\frac{4}{5} - \frac{2}{5} + \frac{3}{5} = \frac{5}{5} = 1$ $\frac{2}{5} + \frac{3}{5} - \frac{4}{5} = \frac{1}{5}$

---

1. I met 3 wolves, 5 pigs, 6 cows, and 4 dogs. What fraction of the animals were pigs?

Answer: five eighteenths.

2. Unscramble each word to find out more about biomes in Minecraft. Write the word in the squares.

retarin | t | e | r | r | a | i | n |

ecahb | b | e | a | c | h |

iatga | t | a | i | g | a |

uaimtonns | m | o | u | n | t | a | i | n | s |

nalsmawpd | s | w | a | m | p | l | a | n | d |

---

1. Use the fractions strips to answer the questions.

$^1/_2$		
$^1/_4$		
$^1/_5$		
$^1/_6$		
$^1/_8$		
$^1/_{10}$		

1in

How many equal parts are in 1 inch (the whole)?

$\frac{1}{2}$?   2 $\frac{1}{6}$?   6

$\frac{1}{4}$?   4 $\frac{1}{8}$?   8

$\frac{1}{5}$?   5 $\frac{1}{10}$?   10

How many equal parts are in $\frac{1}{2}$ of an inch (half of the whole)?

$\frac{1}{4}$? 2 $\frac{1}{6}$? 3 $\frac{1}{8}$? 4 $\frac{1}{10}$? 5

Fill in the missing numerators: $1 = \frac{2}{2} = \frac{10}{10} = \frac{3}{3} = \frac{6}{6} = \frac{5}{5} = \frac{8}{8} = \frac{4}{4}$

How many $\frac{1}{10}$'s are in $\frac{3}{5}$ of an inch? 6

How many $\frac{1}{8}$'s are in $\frac{3}{4}$ of an inch? 6

91

1. Answer the questions.

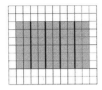

Draw a rectangle in the box and divide it into 5 equal parts. Compare the fractions, using "<", ">", or "=":

$$\frac{5}{5} = 1?\qquad \frac{2}{5} < \frac{4}{5}?$$

2. What fraction name do you need to subtract to get 1 whole?

$$\frac{2}{2} = \frac{3}{3} = \frac{4}{4} = \frac{5}{5} = \frac{6}{6} = \frac{8}{8} = \frac{10}{10} = 1.$$

$$\frac{14}{10} - \frac{4}{10}\qquad \frac{7}{4} - \frac{3}{4}\qquad \frac{9}{6} - \frac{3}{6}\qquad \frac{5}{3} - \frac{2}{3}$$

$$\frac{3}{2} - \frac{1}{2}\qquad \frac{8}{5} - \frac{3}{5}\qquad \frac{10}{8} - \frac{2}{8}$$

3. Add and subtract the fractions.

$$\frac{5}{6} - \frac{1}{6} = \frac{4}{6}\qquad \frac{4}{6} - \frac{3}{6} = \frac{1}{6}\qquad \frac{5}{6} - \frac{3}{6} = \frac{2}{6}$$

$$\frac{5}{6} + \frac{1}{6} - \frac{3}{6} = \frac{3}{6}\qquad \frac{5}{6} - \frac{1}{6} - \frac{2}{6} = \frac{2}{6}\qquad \frac{1}{6} + \frac{5}{6} - \frac{2}{6} = \frac{4}{6}$$

---

1. Let's make an experiment! You need 3 cups, a jar with water, and an empty jar. Fill in the missing fractions or numbers.

1) Fill 1 cup with water. You have a WHOLE cup of water or it equals 1.

2) Take another cup and fill a half of it with water. You have $\frac{1}{2}$ of a cup of water.

3) Pour the first and second cups into the empty jar. You poured 1 cup and $\frac{1}{2}$ of a cup which equals $1\frac{1}{2}$.

1 cup $\frac{1}{2}$ of a cup $1\frac{1}{2}$ of a cup

Compare the 2 cups and the jar and use ">," "<," or "=."

 >   <

Compare the fractions and fill in ">" or "<":

$$1 > \frac{1}{2};\qquad \frac{1}{2} < 1\frac{1}{2};\qquad 1 < 1\frac{1}{2}.$$

1. I planted 2 spruce trees, 5 dark oak trees, and 3 birch trees. What fraction of the trees were birch trees?

Answer: three tenths.

2. Find and circle or cross out the words to find out more about blazes in Minecraft.

```
R M R M O B B Q L F D C      GRAY
Q E I E D J A T S I I N
T L S A N Q E T V R B B      HOSTILE
J J N I Z W A N Y E Y L
V P M F S I A J N B B S      MOB
P E P D R T O P S A S O
Y W U C H F A W S L S T      FIREBALLS
B H A Y C W P N L L J A
Z S S Y T N Y I G S N R      CATCH
E B A J A P C A D E O H
J R C Y C E L I T S O H      RESISTANCE
G C J O L C D E Z I C V
                             STAIRCASE

                             SPAWNER
```

---

1. Write the missing numbers and subtract the fractions.

| 1/6 | | | | |

$$\frac{5}{6} - \frac{2}{6} = \frac{5-2}{6} = \frac{3}{6}\qquad \frac{4}{6} - \frac{3}{6} = \frac{4-3}{6} = \frac{1}{6}$$

| 1/3 | | |

$$\frac{2}{3} - \frac{1}{3} = \frac{2-1}{3} = \frac{1}{3}\qquad \frac{3}{3} - \frac{1}{3} = \frac{3-1}{3} = \frac{2}{3}$$

2. Compare the fractions using ">," "<," or "=."

| 1/3 | | |
| 1/6 | | | | |

$$\frac{2}{6} = \frac{1}{3}\qquad \frac{5}{6} > \frac{2}{3}$$

$$\frac{1}{6} < \frac{1}{3}\qquad \frac{2}{6} < \frac{2}{3}$$

$$\frac{4}{6} = \frac{2}{3}\qquad \frac{3}{3} > \frac{3}{6}$$

1. <u>Add and compare</u> the fractions using ">," "<," or "=.".

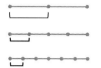

$$\frac{1}{2} + \frac{1}{2} = \frac{1+1}{2} = \frac{2}{2} \qquad > \qquad \frac{1}{4} + \frac{1}{4} = \frac{1+1}{4} = \frac{2}{4}$$

$$\frac{1}{6} + \frac{1}{6} = \frac{1+1}{6} = \frac{2}{6} \qquad < \qquad \frac{2}{6} + \frac{1}{6} = \frac{2+1}{6} = \frac{3}{6}$$

$$\frac{3}{6} + \frac{3}{6} = \frac{3+3}{6} = \frac{6}{6} \qquad = \qquad \frac{1}{4} + \frac{3}{4} = \frac{1+3}{4} = \frac{4}{4}$$

2. <u>Compare</u> the fractions using ">," "<," or "=."

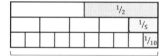

$$\frac{1}{2} = \frac{5}{10}$$

$$\frac{3}{5} = \frac{6}{10}$$

$$\frac{1}{2} > \frac{2}{5} \qquad \frac{4}{5} > \frac{7}{10} \qquad \frac{5}{5} > \frac{9}{10}$$

---

1. <u>Fill in</u> the missing numbers and <u>draw</u> the missing shapes in the box.

 equals $1\frac{3}{4}$ as  equals $1\frac{2}{4}$

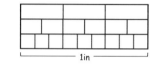 equals $2\frac{1}{2}$ as equals $3\frac{1}{2}$

equals $2\frac{2}{4}$

as equals $4\frac{1}{4}$

2. <u>Compare</u> the fractions.

1in

$$\frac{3}{3} > \frac{4}{6} \qquad\qquad \frac{2}{9} < \frac{2}{6}$$

$$\frac{5}{6} > \frac{5}{9} \qquad\qquad \frac{6}{9} = \frac{4}{6}$$

1. I crafted an axe with ⬜3 oak blocks and ⬜2 sticks. <u>What fraction of an axe is made from sticks?</u>

Answer: two fifths.

---

1. <u>Read</u> and <u>write</u> the missing numbers.

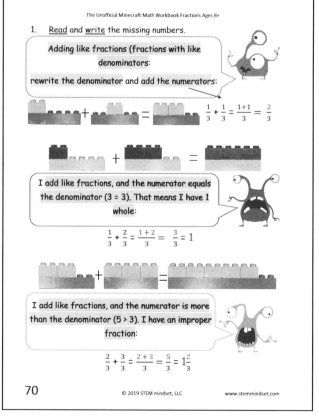

Adding like fractions (fractions with like denominators:

rewrite the denominator and add the numerators:

$$\frac{1}{3} + \frac{1}{3} = \frac{1+1}{3} = \frac{2}{3}$$

I add like fractions, and the numerator equals the denominator (3 = 3). That means I have 1 whole:

$$\frac{1}{3} + \frac{2}{3} = \frac{1+2}{3} = \frac{3}{3} = 1$$

I add like fractions, and the numerator is more than the denominator (5 > 3). I have an improper fraction:

$$\frac{2}{3} + \frac{3}{3} = \frac{2+3}{3} = \frac{5}{3} = 1\frac{2}{3}$$

1. <u>Write</u> the missing numbers.

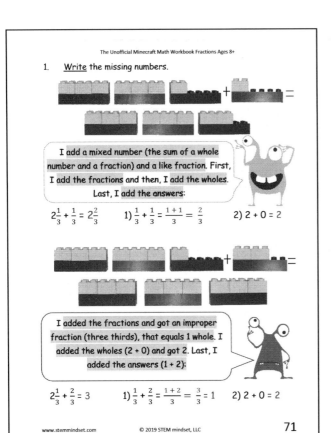

I add a mixed number (the sum of a whole number and a fraction) and a like fraction. First, I add the fractions and then, I add the wholes. Last, I add the answers:

$2\frac{1}{3} + \frac{1}{3} = 2\frac{2}{3}$ 1) $\frac{1}{3} + \frac{1}{3} = \frac{1+1}{3} = \frac{2}{3}$ 2) $2 + 0 = 2$

I added the fractions and got an improper fraction (three thirds), that equals 1 whole. I added the wholes (2 + 0) and got 2. Last, I added the answers (1 + 2):

$2\frac{1}{3} + \frac{2}{3} = 3$ 1) $\frac{1}{3} + \frac{2}{3} = \frac{1+2}{3} = \frac{3}{3} = 1$ 2) $2 + 0 = 2$

---

1. <u>Write</u> the missing numbers.

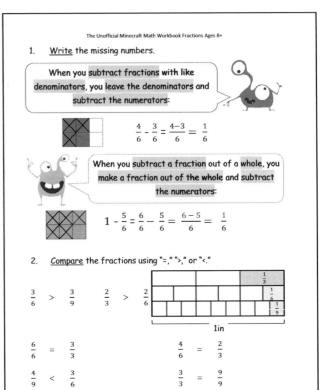

When you subtract fractions with like denominators, you leave the denominators and subtract the numerators:

$\frac{4}{6} - \frac{3}{6} = \frac{4-3}{6} = \frac{1}{6}$

When you subtract a fraction out of a whole, you make a fraction out of the whole and subtract the numerators:

$1 - \frac{5}{6} = \frac{6}{6} - \frac{5}{6} = \frac{6-5}{6} = \frac{1}{6}$

2. <u>Compare</u> the fractions using "=," ">," or "<."

$\frac{3}{6}$ > $\frac{3}{9}$ $\frac{2}{3}$ > $\frac{2}{6}$

$\frac{1}{3}$
$\frac{1}{6}$
$\frac{1}{9}$
1in

$\frac{6}{6}$ = $\frac{3}{3}$ $\frac{4}{6}$ = $\frac{2}{3}$

$\frac{4}{9}$ < $\frac{3}{6}$ $\frac{3}{3}$ = $\frac{9}{9}$

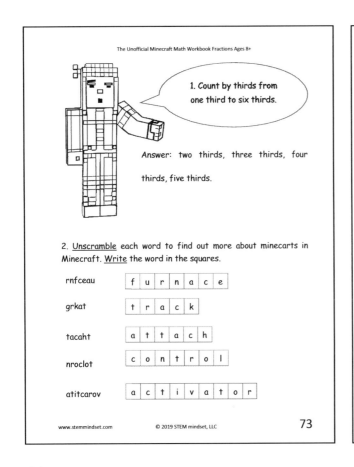

1. Count by thirds from one third to six thirds.

Answer: two thirds, three thirds, four thirds, five thirds.

2. <u>Unscramble</u> each word to find out more about minecarts in Minecraft. <u>Write</u> the word in the squares.

rnfceau | f | u | r | n | a | c | e |

grkat | t | r | a | c | k |

tacaht | a | t | t | a | c | h |

nroclot | c | o | n | t | r | o | l |

atitcarov | a | c | t | i | v | a | t | o | r |

---

1. <u>Find out</u> how long the line segments are. 1 square = $\frac{1}{2}$ of a foot.

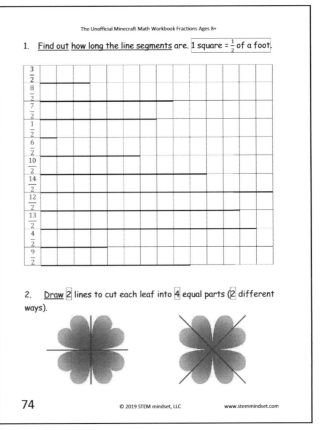

$\frac{3}{2}$
$\frac{8}{2}$
$\frac{7}{2}$
$\frac{1}{2}$
$\frac{6}{2}$
$\frac{10}{2}$
$\frac{14}{2}$
$\frac{12}{2}$
$\frac{13}{2}$
$\frac{4}{2}$
$\frac{9}{2}$

2. <u>Draw</u> 2 lines to cut each leaf into 4 equal parts (2 different ways).

94

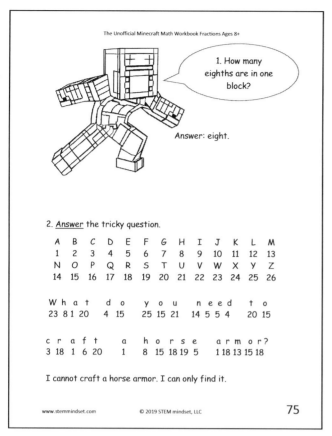

1. How many eighths are in one block?

Answer: eight.

2. <u>Answer</u> the tricky question.

A	B	C	D	E	F	G	H	I	J	K	L	M
1	2	3	4	5	6	7	8	9	10	11	12	13
N	O	P	Q	R	S	T	U	V	W	X	Y	Z
14	15	16	17	18	19	20	21	22	23	24	25	26

W h a t   d o   y o u   n e e d   t o
23 8 1 20   4 15   25 15 21   14 5 5 4   20 15

c r a f t   a   h o r s e   a r m o r?
3 18 1 6 20   1   8 15 18 19 5   1 18 13 15 18

I cannot craft a horse armor. I can only find it.

1.   <u>What fraction names</u> the dark part in each shape below? **Watch out for the tricks!**

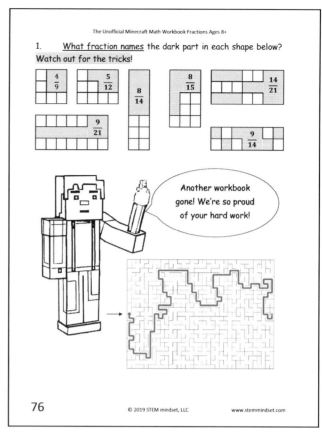

$\frac{4}{9}$ $\frac{5}{12}$ $\frac{8}{14}$ $\frac{8}{15}$ $\frac{14}{21}$ $\frac{9}{21}$ $\frac{9}{14}$

Another workbook gone! We're so proud of your hard work!

Made in the USA
Lexington, KY
17 July 2019